Civil War America

Also by Paul Johnson

Humorists

Jesus: A Biography from a Believer

Churchill

Heroes

Creators

George Washington

Art: A New History

A History of the American People

The Quest for God

The Birth of the Modern

Intellectuals

A History of the English People

A History of the Jews

Modern Times

A History of Christianity

Civil War America

1850–1870

from
A History of the American People
and
Heroes

Paul Johnson

HARPER  PERENNIAL

NEW YORK • LONDON • TORONTO • SYDNEY • NEW DELHI • AUCKLAND

HARPER ● PERENNIAL

"The Almost Chosen People" is excerpted from *A History of the American People,* published in 1997 by Weidenfeld & Nicolson.

"Two Kinds of Nobility" is excerpted from *Heroes,* published in 2007 by HarperCollins Publishers.

HarperCollins books may be purchased for educational, business, or sales promotional use. For information please write: Special Markets Department, HarperCollins Publishers, 10 East 53rd Street, New York, NY 10022.

FIRST EDITION

Designed by Michael P. Correy

Library of Congress Cataloging-in-Publication Data is available upon request.

ISBN 978-0-06-207625-0

11 12 13 14 15 OV/RRD 10 9 8 7 6 5 4 3 2 1

Contents

1 | The Almost Chosen People   1

2 | Two Kinds of Nobility: Lincoln and Lee   147

Source Notes   167

Civil War America

1

The Almost Chosen People

The Civil War, in which are included the causes and consequences, constitutes the central event in American history. It is also America's most characteristic event which brings out all that the United States is, and is not. It made America a nation, which it was not before. For America, as we have seen, was not prescriptive, its people forged together by a forgotten process in the darkness of prehistory, emerging from it already a nation by the time it could record its own doings. It was, rather, an artificial state or series of states, bound together by negotiated agreements and compacts, charters and covenants. It was made by bits of parchment,

bred by lawyers. The early Americans, insofar
as they had a nationality, were English (or more
properly British) with an English national identity and culture. Their contract to become Americans—the Declaration of Independence—did not
in itself make them a nation. On the contrary; the
very word "nation" was cut from it—the Southerners did not like the word. Significantly it was John
Marshall, the supreme federalist, the legal ideolo-
gist of federalism, who first asserted in 1821 that
America was a nation. It is true that Washington
had used the word in his Farewell Address, but el-
liptically, and it was no doubt inserted by Hamil-
ton, the other ideologue of federalism. Washington referred to "the Community of Interest in one
Nation," which seems to beg the question whether
America was a nation or not. And even Marshall's
definition is qualified: "America has chosen to be,"
he laid down, "in many respects and for many
purposes, a nation." This leads one to ask: in what
respects, and for what purposes, was America
not a nation? The word is not to be found in the
Constitution. In the 1820s in the debates over the
"National Road," Senator William Smith of South
Carolina objected to 'this insidious word": he said
it was "a term unknown to the origins and theory
of our government." As one constitutional historian has put it: "In the architecture of nationhood,
the United States has achieved something quite
remarkable . . . Americans erected their constitutional roof before they put up their national walls
. . . and the Constitution became a substitute for a
deeper kind of national identity."[1]

Yes; but whose Constitution: that as seen by the North, or the one which the South treasured—or the one, in the 1850s, interpreted by the Southern-dominated Taney Supreme Court? The North, increasingly driven by emancipationists, thought of the Constitution as a document which, when applied in its spirit, would eventually insure that all people in America, whatever their color, black or white, whatever their status, slave or free, would be equal before the law. The Southerners, by which I mean those who dominated the South politically and controlled its culture and self-expression, had a quite different agenda. They believed the Constitution could be used to extend not so much the fact of slavery—though it could do that too—but its principle. Moreover, they possessed, in the Democratic Party, and in the Taney Court, instruments whereby their view of the Constitution could be made to prevail. They were frustrated in this endeavor by their impetuosity and by their divisions—that is the story of the 1850s.

For the South, the decade began well. True, the California gold rush had been, from their point of view, a stroke of ill-fortune, since the slavery-hating miners who rushed there frustrated the South's plan of making California a slave state. But in some other respects the Compromise of 1850 worked in their favor. For one thing it made it possible for them to keep the Democratic Party united, and since 1828 that party had been the perfect instrument for winning elections. All it had to do, to elect a president of its choosing, was to hold the South together and secure a reasonable slice of the North; then, with their own man in the White House, appointing new Supreme Court judges, they could keep the South's interpretation of the Constitution secure too. For the election of 1852 the Democrats were able to unite round a campaign platform which promised "to abide by and adhere to a faithful execution of the acts known as the Compromise Measures," and for their candidate they picked a man peculiarly adapted to follow that line, "a Northerner with Southern inclinations."

Franklin Pierce (1804–69) was born in Hillboro, New Hampshire, had been to Bowdoin and practiced as a lawyer in Concord. So by rights he should have been an abolitionist and an Emersonian, a political Transcendentalist, and a thorough New Englander. But in reality he was a Jacksonian Democrat, another "Young Hickory" and an ardent nationalist, all-out for further expansion into the crumbling Hispanic South, and thus to that extent a firm ally of the slavery-extenders. He had been a New Hampshire congressman and senator and had served assiduously in the Mexican War, of which (unusual in the North) he was an enthusiastic supporter, reaching the rank of brigadier-general. At the 1852 Democratic convention he emerged, after many votes, as the perfect Dark Horse compromise candidate, being nominated on the forty-ninth ballot. He is usually described as "colorless." When he was nominated, an old farmer-friend from New Hampshire commented: "Frank goes well enough for Concord, but he'll go monstrous thin, spread out over the United States." Nathaniel Hawthorne, who had been a close friend of Pierce at Bowdoin, called on Pierce after he was nominated, sat by him on the sofa, and said: "Frank, what a pity . . . But, after all, this world was not meant to be happy in— only to succeed in." This story is apocryphal, but Hawthorne said something similar to Pierce in a letter in which he undertook to write Pierce's campaign biography. Horace Mann, who knew both, said of the proposed biography, "If he makes out Pierce to be a great man or a brave man, it will be the greatest work of fiction he ever wrote." Hawthorne agreed: "Though the story is true, it took a romancer to do it."[2]

Hawthorne had to conceal two things: Pierce's drinking—it was said he drank even more than Daniel Webster, and he was certainly often drunk— and the fact that he hated Pierce's wife Jane. So did a lot of other people. The Pierces had two sons. Their four-year-old died in 1844; their surviving son was killed a month after the election in an appalling railroad accident, and Jane felt, and said, that the presidency had been bought at the cost of their son's life. Hawthorne burned documents about Pierce which were highly derogatory, commenting: "I wish he had a better wife, or none at all. It is too bad that the nation should be compelled to see such a death's head in the preeminent place among American women; and I think a presidential candidate ought to be scrutinised as well in regard to his wife's social qualifications, as to his own political ones." Jane was the daughter of the Bowdoin president and sister-in-law of its most distinguished professor: but women of academic families are not always congenial.[3] The fact is, Hawthorne hated most women, particularly if they had intellectual pretensions, which Jane certainly did: he said of women writers, "I wish they were forbidden to write on pain of having their faces scarified with an oyster-shell!"[4] At any rate, *The Life of Franklin Pierce* duly appeared, the tale of "A beautiful boy, with blue eyes, light curling hair, and a serene expression of face," who grew up to be a distinguished military man and a conciliatory politician, anxious to preserve the Union by reassuring the South and appealing to "the majority of Northerners" who were "not actively against slavery" to beware of what Hawthorne called "the mistiness of a philanthropic system."[5]

Pierce won handsomely. The Whigs selected the Mexican War commander, General Winfield Scott, who like most generals was lost in the complex politics of ethnic America. He not only bellowed out his anti-slavery views, which the Whigs had allowed for, but often sounded in his speeches a strident nativist only happy with Americans of Anglo-Saxon stock, so he alienated the Germans and the Irish. In the end he carried only Tennessee, Kentucky, Vermont, and Massachusetts, giving Pierce a landslide in the electoral college, though his plurality over all the other candidates (there were four vote-splitters) was only 50,000.[6] In theory Pierce's Cabinet bridged North and South, since his secretary of state, William Learned Marcy (1786–1857), was a member of the old Albany Regency, the New York politico who had egged on Jackson to enjoy "the spoils of victory" in 1829. But Marcy did not care a damn about slavery and, as Polk's secretary of war, had been a rabid architect of the war against Mexico. Again, Pierce's attorney-general, Caleb Cushing (1800–79), though a Harvard–Massachusetts Brahmin, was primarily, like Marcy, a "Manifest Destiny" man, and thus a Southern ally. On the other side, Pierce made Jefferson Davis (1808–89) secretary of war, and Davis was not merely a genuine Southerner but the future president of the Confederation. In practice, then, the Pierce administration was committed to policies which might have been designed to help the South.

The first expression of this policy was the Gadsden Purchase in 1853. This was Davis' idea, significantly. America was then discussing alternative possibilities for transcontinental railways and Davis was determined, for strategic as well as economic reasons, that the South should control one route. This required passage through a large strip of territory in what was then still northwest Mexico. Davis persuaded Pierce to send the South Carolina railroad promoter, Senator James Gadsden (1788–1858), to Mexico to promote the purchase of the strip. This was a dodgy business, as Gadsden had a financial interest in securing the purchase, which was made with U.S. federal money—$10 million for 45,000 square miles—and the Senate agreed to ratify the deal only by a narrow margin, partly because this extra territory automatically became slave soil. Indeed Davis' original idea, that Gadsden should buy not only the strip but the provinces of Tamaulipas, Nuevo Leon, Coahuila, Chihuahua, Sonora, and the whole of Baja (lower) California, was also on the cards but not proceeded with as the Senate knew these vast territories would have been turned into several new slave states, and would never have ratified the deal, the Senate now having a Northern majority, or rather an anti-slave one.[7]

There were other possibilities for the South, however. They wanted Cuba, to turn it into an ideal slave state. "The acquisition of Cuba," wrote Davis, "is essential to our prosperity and security." He regretted that, in joining the Union, the Southern states had forfeited their right to make treaties and acquire new territories on their own, otherwise Cuba would already be in the Union, and slave soil. James Buchanan (1791–1866), who as Polk's secretary of state had been a leading mover in acquiring Texas, was now minister in London and intrigued and negotiated furiously in 1854 to have Cuba purchased and annexed. But nothing came of it—this was one of many occasions when Northerners in Congress frustrated the South's dream of an all-American, all-slave Caribbean.[8] There were various filibustering expeditions to seize by force what might be more difficult to acquire by diplomacy. Prominent in them was William Walker (1824–60), a Tennessee doctor and populist fanatic, who wanted to annex chunks of Latin America to the U.S., not to make them slave states but to give their peoples a taste of democracy. The "gray-eyed man of destiny" entered Lower California in 1853 and proclaimed a republic, but Pierce was not hard-faced enough to allow that. Then Walker took his private army to Nicaragua and actually had himself recognized by the U.S. in 1856. But that aroused the fury of another predator, Cornelius Vanderbilt (1794–1877), whose local transport system was being disrupted by Walker's doings, and as Vanderbilt had more money, he was able to force Walker to "surrender" to the U.S. Navy. Finally Walker turned to Honduras, but there the British navy took a hand and turned him over, as a nuisance, to a Honduran firing-squad.[9]

Now that the Gadsden Purchase made a Southern railway route to California geographically possible, others were looking for northern routes, and this too had an important bearing on the land strategy of the South. Senator Stephen Douglas of Illinois, who had helped Clay to draft the 1850 Compromise, was now chairman of the Senate Committee on Territories, and in that capacity he brought forward a bill to create a new territory called Nebraska in the lands west of the Missouri and Iowa, the object being to get rails across it with an eastern terminus in the rapidly growing beef-and-wheat capital of Chicago. To appease the Southerners, he proposed to include in the bill a popular sovereignty clause, allowing the Nebraskans themselves to decide if they wanted slavery or not. The South was not satisfied with this and Douglas sought to reassure them still further by not only providing for another territory and future state, Kansas, but repealing the old 1820 Missouri Compromise insofar as it banned slavery north of latitude 36.30. This outraged the North, brought up to regard the 1820 Compromise as a "sacred pledge," almost part of the Constitution. It outraged some Southerners too, such as Sam Houston of Texas, who saw that these new territories would mean the expulsion of the Indians, who had been told they could occupy these lands "as long as grass shall grow and water run." But Douglas, who wanted to balance himself carefully between North and South and so become president, pushed on; and President Pierce backed him; and so the Kansas–Nebraska Act passed by 113 to 100 in the House and 37 to 14 in the Senate, in May 1854.[10]

Backing this contentious bill proved, for Pierce, a mistake and ruled out any chance of his being reelected. It also led to what might be called the first bloodshed of the Civil War. Nebraska was so far north that no one seriously believed it could be turned into a series of slave states. Kansas was a different matter, and both sides tried to build up militant colonies there, and take advantage of the new law which stated its people were "perfectly free to form and regulate their domestic institutions in their own way, subject only to the Constitution." The first foray was conducted by the New England Emigrant Aid Society, which in 1855–56 sent in 1,250 anti-slavery enthusiasts. The Southerners organized just across the border in Missouri. In October 1854 the territory's first governor, Andrew H. Reeder, arrived and quickly organized a census, as prelude to an election in March 1855. But when the election

came, the Missourians crossed the border in thousands and swamped the polls. The governor said
the polls were a fraud but did nothing to invalidate
the results, probably because he was afraid of being
lynched. Territorial governors were provided by
Washington with virtually no resources or money,
as readers of Chapter 25 of Mark Twain's *Roughing
It*—which describes the system from bitter experience—will know. At all events the slavers swept the
polls, expelled from the legislature the few anti-slavers who were elected, adopted a drastic slave-code,
and made it a capital offense to help a slave escape
or aid a fugitive. They even made orally question-
ing the legality of slavery a felony.[11]

The anti-slavers, and genuine settlers who
wanted to remain neutral, responded by holding
a constitutional convention—elected unlawfully—
drafted a constitution in Topeka which banned
both slaves and freed blacks from Kansas, applied
for admission to the U.S. as a state, and elected
another governor and legislature. Then the fighting began, a miniature civil war of Kansas' own.
The Bible-thumping clergymen from the North
proved expert gun-runners, especially of what were
known as "Beecher's Bibles," rifles supplied by the
bloodthirsty congregation of the Rev. Henry Ward
Beecher. The South moved in guns too. In May
1856 a mob of slavers sacked Lawrence, a free-soil
town, blew up the Free State Hotel with five cannon, burned the governor's house and tossed the
presses of the local newspaper into the river. This
in turn provoked a fanatical free-soiler called John

Brown, a glaring-eyes fellow later described by one who was with him in Kansas as "a man impressed with the idea that God has raised him up on purpose to break the jaws of the wicked." Two days after the "Sack of Lawrence," Brown, his four sons, and some others rushed into Pottawatomie Creek, a pro-slavery settlement, and slaughtered five men in cold blood. By the end of the year over 200 people had been murdered in "Bleeding Kansas."[12]

The Lawrence outrage in turn provoked a breakdown of law in the Congress. The next day, May 22, Senator Charles Sumner (1811–74) of Massachusetts, a dignified, idealistic, humorless, and golden-tongued man who also had a talent for vicious abuse—the kind which causes wars—delivered a philippic in the Senate. One of the weaknesses of Congressional procedure was that, unlike the British parliament, where a speaker must go on until he finishes, senators were allowed an overnight respite then allowed to start again next morning, provoking their antagonized hearers beyond endurance. In his two-day speech, full of excitable sexual images, Sumner said what was going on in Kansas was "the rape of a virgin territory [sprung] from a depraved longing for a new slave state, the hideous offspring of such a crime." He made a particular target of Senator A. P. Butler of South Carolina, whom he accused of having "chosen a mistress who . . . though polluted in the sight of the world, is chaste in his sight—I mean the harlot, slavery." One cannot help feeling that, in the run-up to the Civil War, sex played a major, if unspoken, part. All Northerners knew, or believed, that male slave-owners slept with their pretty female slaves, and often bought them with this in mind. Abraham Lincoln, aged twenty-two and on his second visit to New Orleans, saw a young and beautiful teenage black girl, "guaranteed a virgin," being sold, the leering auctioneer declaring: "The gentleman who buys her will get good value for his money." The girl was virtually naked, and the horrific scene made a deep impression on the young man. Southerners denied they fornicated with their female slaves, but they also (contradicting themselves) accused their Northern tormenters of sexual envy, which may have been true in some cases.

In any event Sumner's metaphors were provocative. Butler's nephew, Congressman Preston S. Brooks, fumed over the insults for two days, then attacked Sumner with his cane while he was writing at his dcsk in the Senate. Sumner was so badly injured or traumatized, that he was ill at home for two years, his empty Senate desk symbolizing the stop-at-nothing violence of the Southern slavers. Equally significant was that Brooks, having been censured by the House, resigned and was triumphantly reelected, his admirers presenting him with hundreds of canes to mark his "brave gesture," though it was in fact a cowardly assault on an unarmed, older man. Here was a case of unbridled and inflammatory Northern words provoking reckless Southern aggression—a paradigm of the whole conflict.[13]

Brooks' attack, and the support it received from the "gentlemanly South," reflected the aggressive politics of the slave states. The *Dred Scott* verdict by the Taney Court had given the South hope that the constitutional history of the country could be rewritten in a way that would make slavery safe forever. All previous arrangements had left the South insecure—insecurity was at the very root of its violence. What the Southern militants, especially in South Carolina, wanted was a "black code," enacted by Congress and imposed on the territories. They were not so foolish as to hope they could reinstate slavery in New York and New England but they wanted abolitionism to be made illegal in some way. And they wanted not merely to open new territory in the South and West and outside the present borders of the U.S. to slavery but also to reopen and relegalize the slave trade.

This forward plan received an important boost with the election of 1856. The Kansas–Nebraska Act destroyed the last remains of the crumbling Whig Party. In its place, phoenix-like, came the new Republicans, deliberately designed to evoke the memory of Jefferson, now presented as an anti-slaver, his attacks on slavery being eminently quotable, his ownership of slaves forgotten. At its nominating convention, the Republican Party passed over its chief anti-slaver, William H. Seward (1801–72), as too extreme, and picked John Charles Frémont (1813–90), a South Carolina adventurer who had eloped with the daughter of old Senator Benton and then had innumerable near-death escapes in California, including a capital conviction for mutiny quashed by President Polk. The Repub-

lican slogan was "Free Soil, Free Speech and Frémont." The Democratic Party, rejecting Pierce as a
sure loser, and Douglas as too all-things-to-all-men,
picked James Buchanan, who concentrated on taking
all the slave states and as much of the rest as he could.
Old Fillmore, with Jackson's son-in-law Donelson as
his running mate, popped up from the past as a
splitter. That did it for Frémont. So Buchanan, with
a fairly united Democratic Party behind him, carried all the South plus New Jersey, Pennsylvania,
Illinois, Indiana, and California, making 174 college voters, against Frémont's 114. Buchanan was
elected on a minority (45.3 percent) of the vote but
his plurality over Frémont was wide, 1,838,169 to
1,341,264.

The new president was at heart a weak man,
and a vacillating one, but he was not out of touch
with the combination of imperialist and Southern
opinion which, well led, would have ruled out any
prospect of coercion of the South by the North.
Whatever he said in public, Buchanan sympathized
with the idea of adding new states to the South,
even if slavers. In his message to Congress, January 7, 1858, Buchanan criticized Walker's filibustering in Nicaragua not because it was wrong in
itself but because it was impolitic and "impeded the
destiny of our race to spread itself over the continent of North America, and this at no distant day,
should events be permitted to take their natural
course." He followed this up by asking Congress to
buy Cuba, despite the fact that the Spanish were
demanding at least $150 million for it (the Republi-

cans blocked the plan). America had absorbed what was once Spanish-speaking territory of millions of square miles in California and Texas: why not the whole of Mexico and Central America? That was all part of the "North American Continent," to which the U.S. was "providentially entitled" by its Manifest Destiny.[14]

Moreover, the price of slaves was rising all the time, despite the efforts of the Virginia slave-farms to produce more, and this in turn strengthened demands for a resumption of the slave-trade. Slave-smuggling was growing, and it was well known, and trumpeted in the South, that merchants in New York and Baltimore bought slaves cheap on the West African coast, and then landed them on islands off Georgia and other Southern states. So why not repeal the 1807 Act and legalize the traffic? That was the demand of the governor of South Carolina in 1856, and the Vicksburg Commercial Convention of 1859 approved a motion resolving that "all laws, state or federal, prohibiting the African slave trade, ought to be repealed." The first step, it was argued, was to have blacks captured from slave-ships stopped and searched by the U.S. Navy—the current practice was to send them, free, to Liberia, which most of them did not like—sent to the South and "apprenticed" to planters with good records. Representative William L. Yancey of Alabama asked: "If it is right to buy slaves in Virginia and carry them to New Orleans, why is it not right to buy them in Cuba, Brazil or Africa, and carry them there?" If blacks would rather be slaves in the South than free men in Liberia, might it not be that other African blacks would prefer to come to the South, as slaves, rather than remain in the "Dark Continent," where their lives were so short and cheap?

Southerners argued that to take a black from Africa and set him up in comfort on a plantation was the equivalent, allowing for racial differences, of allowing a penniless European peasant free entry and allowing him, in a few years, to buy his own farm. The *Dred Scott Case*, by declaring the Missouri Compromise unconstitutional, and the Kansas–Nebraska Act together opened up enormous new opportunities for setting up slave-plantations and ranches, and therefore increased the demand for slaves. Southerners argued that by resuming the slave-trade the cost of slaves in America would be sharply reduced, thereby boosting the economy of the whole country. The aggressive message of the South was: slavery must be extended because it makes economic sense for America. But beneath this aggressive tone was the deep insecurity of Southerners who had no real moral answer to the North's case and knew in their hearts that the days of slavery were numbered.[15]

That sense of insecurity was justified, because in the late 1850s it became obvious that dreams of a vast expansion of slavery to the west and into the Caribbean and other Hispanic areas were fantasies, and the reality was a built-in and continuing decline of Southern political power. Calhoun, in almost his dying words in 1850, had warned the South that if they did not act soon, and assert his theory of states' rights, if necessary by force, they were doomed to a slow death: they would never be stronger than they were, and could only get weaker. That was demonstrated to be good advice; in May 1858 the free state of Minnesota entered the Union, followed by another free state, Oregon, in February 1859, while Kansas, being a slave territory, was denied admission. So the Congressional balance, as Calhoun had foreseen, was destroyed forever. The South was now outvoted in the Senate 36 to 30 and in the House the gap was enormous, 147 to 90.

Southerners' sense of insecurity was deepened by the fact that, while they boasted publicly that "Cotton is King" and "The Greatest Staple in the World," they were painfully aware of the weaknesses of their cotton-slave economy. Most plantations were in debt or operated close to the margins of profitability. During the 1850s, world cotton prices tended to fall. More and more countries were producing raw cotton—a trend which would knock large nails in the South's coffin when the war began. In the light of economic hindsight, it can be seen that the plantation system, as practiced, was fundamentally unsound, and some planters grasped this at the time. Plantations absorbed good land and ruined it, then their owners moved on. There was an internal conflict in the South, as the newer estates in the Deep South were more scientific and efficient (and bigger), and thus tended to take black slave labor away from the tidewater and border areas, and push up the price of slaves. This, at a time of falling cotton prices, put further pressure on profit margins.[16] As the price of slaves rose, slavery as an institution became more vital to the South: to the Deep South because they used slaves more and more efficiently, to the Old and border South because breeding high-quality, high-priced slaves was now far more important than raising tobacco or cotton. Professor Thomas R. Dew of William and Mary College, in his book *The Pro-Slavery Argument* of 1852, asserted: "Virginia is a *negro-raising* state for other states: she produces enough for her own supply and 6,000 [annually] for sale."

Actually, Virginia was living on its slave-capital: blacks formed 50 percent of the Virginia population in 1782, but only 37 percent in 1860s—it was selling its blacks to the Deep South. Virginia and other Old South or border states concentrated on breeding a specially hardy type of negro, long-living, prolific, disease-free, muscular, and energetic. In the 1850s, about 25,000 of these blacks were being sold, annually, to the Deep South.[17] The 1860 census showed there were 8,099,000 whites in the South and 3,953,580 slaves. But only 384,000 whites owned the slaves: 10,781 owned fifty and more; 1,733, a hundred and more. So over 6 million Southern whites had no direct interest in slavery. But that did not mean they did not wish to retain the institution—on the contrary: poor whites feared blacks even more than the rich ones did. By 1860 there were already 262,000 free blacks in the Southern states, competing with poor whites for scarce jobs, and a further 3,018 were manumitted that year. Poor whites were keener than anyone on penal legislation against slaves: they insured no state recognized slave marriage in law, and five states made it unlawful to teach slaves to read and write. In any event, small white farmers in the South were very much at the mercy of the big plantation owners and had to go along with them.[18] Those who produced cotton, rice, sugar, tobacco, and slaves on a large scale were all-powerful. As one historian has put it, "There was never in America a more perfect oligarchy of businessmen."[19]

Slavery was not the only issue between North and South. Indeed it is possible that an attempt at secession might have been made even if the slavery issue had been resolved. The North favored high tariffs, the South low ones; the North, in consequence, backed indirect taxation, the South direct taxation. It is significant that once the war began, the North, shorn of the South, immediately introduced high tariffs with the Morrill Act of 1861, and pushed through direct federal income tax too. There were huge differences of interest over railroad strategy. Increasingly, the railroad interests of the Northeast and the Northwest came into alignment in the 1850s, and this in turn led to an alliance between Eastern manufacturers seeking high tariffs and Western farmers demanding low-cost or free lands—both linked by lines of rail. This was the basis of the power of the new Republican Party, and the South saw it as a plot—indeed, it was what finished them. Many Southerners believed deeply in their hearts that the moral indignation of the North was spurious, masking meaner economic motives. As Jefferson Davis put it, "You free-soil agitators are not interested in slavery . . . not at all . . . It is so that you may have an opportunity of cheating us that you want to limit slave territory within circumscribed bounds. It is so that you may have a majority in the Congress of the United States and convert the government into an engine of Northern aggrandisement . . . you desire to weaken the political power of the Southern states. And why? Because you want, by an unjust system of legislation, to promote the industry of the North-East states, at the expense of the people of the South and their industry."[20]

Davis was reflecting a bitter conviction held by all "thinking" men in the South; that the North, while accusing the South of exploiting the blacks, exploited the whole of the South systematically and without mercy. Their feeling was exactly the same as the resentment felt by the Third World towards the First World today. There was something inherent in a plantation economy which put it in a dependent position, with the capitalist world its master. There was, of course, no control by the state of national production and prices, of cotton or anything else. If world markets were high, profits rose, but there was then a tendency to reinvest them in increased production. If prices fell, the planters had to borrow. In either case, the South lacked liquid capital. So the planters fell into the hands of bankers, ending up dependent on New York or even the City of London.[21] The South lacked its own financial system, like the Third World today. When cotton made big

profits, it spent them, as the Arab rulers today dissi-
pate colossal oil revenues. And it was in a real sense
milked, like the primary producers today in Africa
and Latin America, at the same time accumulating
massive debts it had no hope of repaying. In effect,
the South had all the disadvantages of a one-crop
economy. It had only 8 percent of U.S. manufac-
tures. It should have put up the money to open factories, and so provide employment for poor whites
and diversify its economy at the same time. But
there was no spare capital in the South itself, and
the North had no intention of building factories
there and competing against itself with low-wage,
low-price products. So the South saw itself as the
slave of a Union dominated by Northern capital.
As the *Charleston Mercury* put it: "As long as we are
tributaries, dependent on foreign labor and skill for
food, clothing and countless necessities of life, *we
are in thralldom.*"[22]

The Civil War was not only the most characteristic event in American history, it was also the most
characteristic religious event because both sides were
filled with moral righteousness for their own cause
and moral detestation of the attitudes of their opponents. And the leaders on both sides were righteous
men. Let us look more closely at these two paladins,
Abraham Lincoln and Jefferson Davis. Lincoln
was a case of American exceptionalism because, in
his humble, untaught way, he was a kind of moral
genius, such as is seldom seen in life and hardly ever
at the summit of politics. By comparison, Davis was

a mere mortal. But, according to his lights, he was a just man, unusually so, and we can be confident that, had he and Lincoln been joined in moral discussion, with the topic of slavery alone banned, they would have found much common ground.

Both men were also characteristic human products of mid–19th-century America, though their backgrounds were different in important respects. Lincoln insisted he came from nowhere. He told his campaign biographer, John Locke Scripps of the *Chicago Tribune,* that his early life could be "condensed into a single sentence from Gray's *Elegy,* 'The short and simple annals of the poor.'" He said both his parents were born in Virginia and he believed one of his grandfathers was "a Southern gentleman." He also believed his mother was illegitimate, probably rightly. He was born in a log cabin in the Kentucky backwoods and grew up on frontier farms as his family moved westwards. His father was barely literate; his mother taught him to read, but she died when he was nine. Thereafter he was self-taught. His father remarried, then took to hiring out his tall lanky (six feet four and 170 pounds) son, for 25 cents a day. He said of his son: "He looked as he had been rough-hewn with an axe and needed smoothing down with a jackplane." Lincoln acquired, in the backwoods of Kentucky, Indiana, and Illinois, and on the Ohio and the Mississippi, an immense range of skills: rafting, boating, carpentry, butchering, forestry, store-keeping, brewing, distilling, plowing. He did not smoke, chew tobacco, or drink. He acquired an English grammar, and taught it to himself. He read Gibbon, *Robinson Crusoe,* Aesop, *The Pilgrim's Progress,* and Parson Weems' lives of Washington and Franklin. He learned the *Statutes of Illinois* by heart. He rafted down to New Orleans and worked his way back on a steamer. He visited the South several times and knew it, unlike most Northerners.[23] He listened often to Southerners defend the "Peculiar Institution" and knew their arguments backwards; what he had personally witnessed made him reject them, utterly, though he never made the mistake of thinking them insincere or superficial. He loved Jefferson, Clay, and Webster, in that order. He was a born storyteller, a real genius when it came to telling a tale, short or long. He knew when to pause, when to hurry, when to stop. He was the greatest coiner of one-liners in American history, until Ronald Reagan emerged to cap him. He was awkward—he always put his whole foot flat down when walking, and lifted it up the same way—but could suddenly appear as if transfigured, full of elegance. With one hand he could lift a barrel of whiskey from floor to counter. He was hypochondriac, as he admitted. He wrote an essay on suicide. He said: "I may seem to enjoy life rapturously when I am in company. But when I am alone I am so often so overcome by mental depression that I never dare carry a penknife."[24]

Lincoln was a self-taught lawyer but his instincts were not for the cause. He said, "Persuade your neighbors to compromise whenever you can . . . As a peacemaker, the lawyer has a superior opportunity of being a good man. There will still be business enough. A worse man can scarcely be found than one who [creates litigation]." As a circuit lawyer, Lincoln fancied himself a Whig and stood for the state legislature. His first elective post, however, was as captain of volunteers in the Black Hawk War (1832), in which he came across five scalped corpses in the early morning: "They lay heads towards us on the ground. Every man had a round red spot on the top of his head about as big as a dollar where the redskins had taken his scalp. It was frightful. But it was grotesque. And the red sunlight seemed to paint everything over." But he held no grudge; indeed he saved an Indian from being butchered. He was the first man to refer to Indians as "Native Americans," though in the then current usage the term referred to Americans of old Anglo-Saxon stock. He said to those who protested about German immigrants, and claimed the title for themselves: "Who are the [real] Native Americans? Do they not wear the breechclout and carry the tomahawk? We pushed them from their homes and now turn on others not fortunate enough to come over so early as we or our forefathers."[25]

He did not win his first political election. And he had bad luck. He bought a store and set up as postmaster too. His partner, Berry, fled with the cash and Lincoln had to shoulder a $1,100 burden of debt. Like Washington, he went into land-surveying to help pay it off. Then he was elected to the state assembly, serving eight years from the age of twenty-five to thirty-two. It met in Vandalia, its eighty-three members being divided into two chambers. Lincoln was paid $3 for each sitting, plus pen, ink, and paper. His first manifesto read: "I go for all sharing the privileges of government who assist in sharing its burdens. Consequently I go for admitting all whites to the right of suffrage who pay taxes or bear arms (by no means excluding females)." He belonged to a group of Whig legislators who were all six feet or over, known as the Long Nine. He got the state capital shifted to Springfield and there set up a law practice, making his name by winning a case for an oppressed widow. A colleague said: "Lincoln was the most uncouth-looking man I ever saw. He seemed to have but little to say, seemed to feel timid, with a tinge of sadness visible in his countenance. But when he did talk all this disappeared for the time, and he demonstrated he was both strong and acute. He surprised us more and more at every visit."[26]

Lincoln's first love, Ann Rutledge, died of typhoid fever. That Lincoln was devastated is obvious enough; that his love for her persisted and prevented him from loving any other woman is more debatable.[27] At all events, it is clear he never loved the woman he married, Mary Todd. She came from a grand family in Kentucky, famous since Revolutionary days for generals and governors. She was driven from it by a horrible stepmother, but never abandoned her quest for a man she could marry in order to make him president. Oddly enough, she turned down Stephen Douglas, then a youngish fellow-member of the Illinois Assembly, in favor of Lincoln, whom she picked out as White House timber. She said to friends: "Mr Lincoln is to be president of the United States some day. If I had not thought so, I would not have married him, for you can see he is not pretty." Lincoln consented, but missed the wedding owing to an illness which was clearly psychosomatic. This led to a sabre duel with Sheilds, the state auditor, which was called off when Lincoln scared his opponent by cutting a twig high up a tree. And this in turn led to reconciliation with Mary, and marriage, he being thirty-three, she twenty-four. His law partner, William H. Herndon, said: "He knew he did not love her, but he had promised to marry her."

It was an uncomfortable marriage of opposites, particularly since she had no sense of humor, his strongest suit. He liked to say: "Come in, my wife will be down as soon as she gets her trotting-harness on." He was a messy man, disorderly in appearance, she was a duster and polisher and tidier. She wrangled acrimoniously with her uppity white servants and sighed noisily for her "delightful niggers." "One thing is certain," she said, "if Mr Lincoln should happen to die, his spirit will never find me living outside the boundaries of a slave state." She hated his partner, his family, and his so-called office. Herndon said: "He had no system, no order; he did not keep a clerk; he had neither library, nor index, nor cash-book. When he made notes, he would throw them into a drawer, put them in his vest-pocket, or into his hat . . . But in the inner man, symmetry and method prevailed. He did not need an orderly office, did not need pen and ink, because his workshop was in his head."

The Lincolns had four sons. Generations of Lincoln-admirers have played down the role of Mary in his life and career, easily finding spicy material illustrative of her shortcomings. But the likelihood is that he would never have become president without her. It took him four years, aged thirty-three to thirty-seven, to get into Congress, and but for her endless pushing he might have become discouraged. For his part, he did his best to behave to her gallantly. There is a touching photograph of her, taken in 1861, arrayed in her inaugural finery, wearing pearls. They were a set which Lincoln had just bought for her, paying $530, at Tiffany's store on 550 Broadway: a seed-pearl necklace and matching bracelets for each arm. They are now in the Library of Congress.[28]

Lincoln won a seat for Congress in 1847, by a big majority. The Whig Party gave him $200 for his expenses. He handed back $199.25, having bought only one barrel of cider. He rode to Washington on his own horse, and stayed with friends. But he served only one term—his opposition to the Mexican War determined that for him. He recalled that at the foot of the Capitol, within sight of its windows, was "a sort of negro stable where gangs of negroes were sold, and sometimes kept in store for a time pending transport to the Southern market, just like horses." Lincoln was broad-minded, tolerant, and inclined to let things alone if possible, but he found this insult to the eye of freedom, literally within sight of Congress, "mighty offensive." The first law he drafted was a Bill to Abolish Slavery in the District of Columbia, to be enacted by local referendum (as we have seen it became part of the 1850 Compromise). At the end of his term, he returned contentedly to the law.[29]

But the slavery issue would not let him rest, or stay out of politics. It was even more persistent than Mary Lincoln's pushing. Some notes have survived of his musings:

> If A can prove, however conclusively, that he may, of right, enslave B, why may not B snatch the same argument, even prove equally, that he may enslave A? You say A is white and B is black—is it *color* then, the lighter having the right to enslave the darker? Take care—by this rule, you are to be slave to the first man you meet, with a fairer skin than your own. You do not mean *color* exactly? You mean the whites are *intellectually* the superior of the blacks, and therefore have the right to enslave them? Take care again—by this rule you are to be the slave of the first man you meet, with an intellect superior to your own.[30]

As Herndon said, "All his great qualities were swayed by the despotism of his logic." There are many memorable descriptions of him lost in thought, turning things over in his mind.

Lincoln did a lot of this musing at home, a place in which he kept a low profile. Mary Lincoln said: "He is of no account when he is at home. He never does anything except to warm himself and read. He never went to market in his life. I have to look after all that. He just does nothing. He is the most useless, good-for-nothing man on earth." He replied, in his own way: "For God, one 'd' is enough, but the Todds need two." He was often driven from his own house by Mary's anger. There are no fewer than six eyewitness descriptions of her furies, one relating to how she drove him out with a broomstick. He was never allowed to ask people to a meal, even or rather especially his parents. He wrote: "Quarrel not at all. No man resolved to make the most of himself can spare time for personal contention . . . Yield larger things to which you can show no more than equal right; and yield lesser ones, though clearly your own." Mary felt his righteousness as well as his awkwardness: "He was mild in his manner," she said, "but a terrible firm man when he set his foot down. I could always tell when, in deciding anything, he had reached his ultimatum. At first he was very cheerful, then he lapsed into thoughtfulness, bringing his lips together in a firm compression. When these symptoms developed, I fashioned myself accordingly, and so did all others have to do, sooner or later."[31]

That Lincoln, as his wife implied, had a huge will when intellectually roused to a moral cause is clear. This sprang from a compulsive sense of duty rather than ambition as such. The evidence suggests that he was obliged to reenter politics not because he was an anti-slavery campaigner but because, in the second half of the 1850s, the slavery issue came to dominate American politics to the exclusion of almost everything else. Each time the issue was raised, and Lincoln was obliged to ponder it, the more convinced he became that the United States was uniquely threatened by the evil, and its political consequences. In those circumstances, an American who felt he had powers—and Lincoln was conscious of great powers—had an inescapable duty to use them in the Union's defense. Lincoln did not see slavery in religious terms, as the "organic sin" of the Union, as the Protestant campaigners of the North put it. Those close to him agreed he had no religious beliefs in the conventional sense. His wife said: "Mr Lincoln had no faith and no hope in the usual acceptation of those words. He never joined a church. But still, I believe, he was a religious man by nature . . . it was a kind of poetry in his nature." Herndon said Lincoln insisted no personal God existed and when he used the word God he meant providence: he believed in predestination and inevitability.[32]

Lincoln came closer to belief in God, as we shall see, but in the 1850s he was opposed to slavery primarily on humanitarian grounds, as an affront to man's natural dignity; and this could be caused by religious sectarians as well as by slave-owners. In his boyish and youthful reading, he had conceived great hopes of the United States, which he now feared for. He wrote: "Our progress in degeneracy appears to me to be pretty rapid. As a nation we began by declaring that 'all men are created equal.' We now practically read of 'all men are created equal except negroes.' When the Know-Nothings get control it will be 'all men are created equal except negroes and foreigners and Catholics.'

When it comes to this, I shall prefer emigration to some country where they make no pretence of loving liberty—to Russia, for instance, where despotism can be taken pure, without the base alloy of hypocrisy."[33] The state of America caused him anguish. He said to Herndon: "How hard it is to die and leave one's country no better than if one had never lived for it! The world is dead to hope, deaf to its own death-struggle. One made known by a universal cry, what is to be done? Is anything to be done? Who can do anything? And how is it to be done? Do you never think of these things?"[34]

But from this general sense of downward moral plunging, which had to be arrested, the slavery issue, and still more the South's determination to extend and fortify it, loomed ever larger. In an important letter to Joshua F. Speed, the storekeeper with whom he shared some of his most intimate thoughts, Lincoln dismissed the claim that slavery was the South's affair and Northerners "had no interest" in the matter. There were, he said, many parts of the North, in Ohio for instance, "where you cannot avoid seeing such sights as slaves in chains, being carried to miserable destinations, and the heart is wrung. It is not fair for you to assume that I have no interest in a thing which has and continually exercises, the power of making me miserable." Lincoln was as much concerned for the slave-owner as for the slave—the institution morally destroyed the man supposed to benefit from it. It was thus more important, as Lincoln saw it, to end slave-owning than to end slavery itself. He said a Kentuckian had once told him: "You might have any amount of land, money in your pocket, or bank stock, and while traveling around nobody would be any the wiser. But if you have a darky trudging at your heels, everybody would see him and know you owned a slave. It is the most glittering property in the world. If a young man goes courting, the only inquiry is how many negroes he, or she, owns. Slave-ownership betokens not only the possession of wealth but indicates the gentleman of leisure, who is above labor and scorns it."[35] This image of the strutting slave-owner, corrupted and destroyed by the wretch at his heels, haunted Lincoln. He wept for the South in its self-inflicted moral degradation.

It was because slavery made him miserable, and because he thought it was destroying the nation, not least the South, that Lincoln reentered politics and helped to create the new Republican Party, primarily to prevent slavery's extension. Looking back with the hindsight of history, we tend to assume that slavery was a lost cause from the start and the destruction of the old South inevitable. But to a man of Lincoln's generation, the South appeared to have won all the political battles, and all the legal ones. So long as the Democratic Party remained united, the South's negative grip on the United States seemed unbreakable, and its power to make positive moves was huge. The creation of the Republican Party, from free-soilers, Whigs, and many local elements, was the answer to the Democratic stranglehold on the nation, which had been the central fact of American political life since 1828.

Lincoln failed to get into the Senate in 1855 and (as we have seen) Buchanan won the presidency in 1856. But it was by then apparent that the Republican Party was a potential governing instrument, and Lincoln's part in creating it was obvious and recognized.

At Bloomington on May 29, 1856, when the new Illinois Republican Party was inaugurated, Lincoln was called to make the adjournment speech and he responded with what all agreed was the best speech of his life. It was so mesmerizing that many reporters forgot to take it down. Even Herndon, who always took notes, gave up after fifteen minutes and "threw pen and paper away and lived only in the inspiration of the hour."[36] Lincoln argued that the logic of the South's case, which was that slavery was good for the negroes, would be to extend it to white men too. Because of the relentless pressure of the South's arguments, Northerners like Douglas, Lincoln warned, were now yielding their case of "the individual rights of man"— "such is the progress of our national democracy." Lincoln said it was therefore urgent that there should be a union of all men, of whatever politics, who opposed the expansion of slavery, and said he was "ready to fuse with anyone who would unite with him to oppose slave power." If the united opposition of the North caused the South "to raise the bugbear of disunion," the South should be told bluntly, *the union must be preserved in the purity of its principles as well as in the integrity of its territorial parts.*" And he updated the reply of Daniel Webster to the South Carolina nullifiers, as the

slogan of the new Republican Party: "Liberty and
Union, now and forever, one and inseparable."[37]
One eyewitness said: "At this moment, he looked to
me the handsomest man I had ever seen in my life."
Herndon recalled: "His speech was full of fire and
energy and force. It was logic. It was pathos. It was
enthusiasm. It was justice, equity, truth and right
set alight by the divine fires of a soul maddened
by the wrong. It was hard, heavy, knotty, gnarly,
backed with wrath."[38]

It was now only a matter of time before Lincoln
became the champion of the new Republicans. The
Senatorial election of 1858 in Illinois, when he was
pitted against Douglas, the "Little Giant," provided
the opportunity. On June 16 Lincoln, having been
nominated as Republican candidate, laid down
the strategy at the state convention in Springfield.
Together with the Bloomington speech, it repre-
sents the essence of Lincoln's whole approach to the
complex of political issues which revolved round
slavery. He said that all attempts to end both the
South's agitation for the right to extend slavery and
the North's to abolish it had failed, and that the
country was inevitably moving into crisis:

> A house divided against itself cannot stand. I be-
> lieve this government cannot endure half *slave*
> and half *free*. I do not expect the Union to be *dis-
> solved*. I do not expect the House to *fall*. But I *do*
> expect it will cease to be divided. It will become
> *all* one thing, or *all* the other. Either the *opponents*

of slavery will arrest the further spread of it, and place it where the public mind shall rest in the belief that it is in course of ultimate extinction; or its *advocates* will push it forward, till it shall become alike lawful in *all* the states, *old* as well as *new, North* as well as *South.* [Emphasis Lincoln's.]

The burden of the speech was a masterly summary of the legal and constitutional threats represented by the *Dred Scott* decisions and the Kansas-Nebraska Act, and Lincoln challenged Douglas—his main opponent in the state—to say clearly where he stood on both these issues. Lincoln said of his speech: "If I had to draw a pen across my record, and erase my whole life from sight, and if I had one poor gift or choice left as to what I should save from the wreck, I should choose that speech and leave it to the world unerased."[39]

Lincoln was right to put his finger on Douglas, for he represented the spirit of compromise where it was no longer possible—where further attempts to evade the dread issue would play into the hands of the South and sell the pass. Lincoln objected strongly to Horace Greeley's plan to get Douglas into the Republican Party. He saw Douglas as an unprincipled man motivated solely by ambition. Eventually both North and South came round to Lincoln's view. But in 1858 Douglas was a much weightier politician than Lincoln, albeit a younger man. Only five feet high, but muscular and stocky, he was the son of a doctor but had done many things—laborer in his teens, a teacher at twenty, a lawyer at twenty-one, a state legislator and secretary of state of Illinois, a judge of its supreme court, then a congressman, a senator before he was forty, a European traveler who had been received by the Tsar of Russia and the Queen of England, a rich man who had married two heiresses in turn. He traveled in princely fashion, by special train or coach, with a truck and field gun behind, which fired a salute when he arrived in any place he was due to speak. He drove to his engagements in a carriage with six horses and with thirty-two outriders. So Douglas was a grand man who looked down his nose at the uncouth Lincoln. But Lincoln was cunning when he wished to be. Annoyed by the conservative *Springfield Journal,* he persuaded it to publish an apology for Southern slavery and so ruined its reputation among right-thinking Illinois readers—it went out of business. Determined to get maximum publicity for his House Divided strategy, he provoked and teased and inveigled Douglas into giving him a series of public debates, from which Lincoln had everything to gain and very little to lose.

The Lincoln–Douglas debates were a series of seven encounters, August–October 1858, conducted throughout the state, with the Senate seat the prize. They were preceded and followed by bands and processions and attracted crowds of 10,000 or more, entire families traveling up to 30 miles to attend them. Both men were good debaters and they made a striking contrast of style, Douglas, meticulously dressed, exuding vigor, Lincoln shambling and awkward in word and gesture, then suddenly, without warning and for brief seconds, becoming godlike in his majestic passion. Douglas won the seat. But the debates eventually finished him, while they transformed Lincoln into a national figure. They were, also, an important process in educating the North in the real issues at stake, and this was of far greater historical importance than the Clay–Webster–Calhoun encounters of 1850.[40]

The strength of Douglas was his warning that the path Lincoln was treading could lead to sectional discord on a scale the country had never known, and possibly civil war. His weakness was that he was never really prepared to say where he stood on slavery and was thus exposed, in debate, as trying to be all things to all men. He said: "I do not care whether the vote goes on for or against slavery. That is only a question of dollars and cents. The Almighty himself has drawn across this continent a line on one side of which the earth must be forever tilled by slave labor, whereas on the other side of that line labor is free." Northerners might accept this—indeed had always accepted it—as a convenient or inescapable fact—but they did not want it spelled out. To do so sounded amoral or even immoral. And most Americans, then as now, wanted to sound moral. Then again, Douglas said: "When the struggle is between the white man and the negro, I am for the white man. When it is between the negro and the crocodile, I am for the negro." That too played into Lincoln's hands: it was a remark which would do for a saloon but not for a public platform. Lincoln rightly saw that the debate, the entire controversy, had to be conducted on the highest moral plane because it was only there that the case for freedom and Union became unassailable. He pointed out again and again that even the South was, in its heart, aware that slavery was wrong. The United States had made it a capital offense half a century ago to import slaves from Africa, and that fact, over the years, had wormed its way into Southern attitudes, however much they might try to defend slavery. Hence, even in the South, the slave-dealer was treated with abhorrence. Slave-owners would not let their children play with his—though they would cheerfully see them playing with slave-children. And the South knew that not only slave-dealing was wrong but slavery itself— why else did they manumit: "Why have so many slaves been set free, except by the promptings of conscience?" As for the *Dred Scott* decision, it was an aberration, which would shortly be set right, at the next presidential election: "You can fool all the people some of the time, and some of the people all the time, but you cannot fool all of the people all of the time."[41]

Lincoln's object was not merely to put his name and his case before the American people, as well as Illinois voters. It was also to expose the essential pantomime-horse approach of a man who tried to straddle North and South. He succeeded in both. He put to Douglas the key question: "Can the people of a United States territory, in any lawful way, against the wish of a citizen of the United States, exclude slavery from its limits prior to the formation of a state constitution?" If Douglas said yes, to win Illinois voters, he lost the South. If he said no, to win the South, he lost Illinois. Douglas' answer was: "It matters not what way the Supreme Court may hereafter decide as to the abstract question whether slavery may or may not go into a territory under the Constitution; the people have the lawful means to introduce it or exclude it as they please, for the reason that slavery cannot exist a day or an hour unless it is supported by the local police regulations." This answer won Douglas Illinois but it lost him the South and hence, two years later, the presidency.[42] Lincoln, normally a generous and forgiving man, had no time for Douglas and did not regret destroying his future career. He thought less of Douglas than he did of the Southern leaders. He said: "He is a man with tens of thousands of blind followers. It is my business to make some of those blind followers see."

The debates gave Lincoln precisely the impetus he needed. He quoted Clay many times and in a way he inherited Clay's mantle. The rhyme went: "Westward the star of empire takes its way— the girls link onto Lincoln, their mothers were for Clay." He was told: "You are like Byron, who woke to find himself famous." By 1859 he knew he ought to be president, wanted to be president, and would be president. The campaign autobiography he wrote December 20, 1859 is brief (800 words), plain, and self-dismissive, yet it exudes a certain confidence in himself and his purpose. He sums up his bid for the presidency in two laconic sentences: "I was losing interest in politics, when the repeal of the Missouri Compromise aroused me again. What I have done since then is pretty well known."[43] William Henry Seward (1801–72) and Salmon Portland Chase, (1808–73) were both initially considered stronger contenders for the Republican nomination than Lincoln. Seward, first governor then senator for New York, was the leader of the abolitionists, who said he was "guided by a higher law than the Constitution." Chase was senator, then governor of Ohio, a free-soiler and Democrat who drafted the first Republican Party set of beliefs.[44] Both had strong claims but Lincoln had a big success in New York. At the Republican State convention in Decatur, Lincoln's cousin John Hanks did a remarkable if unconscious public relations job by holding a demonstration centered around two fence-rails which, he said, were among the 3,000 Lincoln had split thirty years before. He told stories of Lincoln's youth and his pioneering father—entirely fanciful in the latter's case—and made rail-splitting into a national symbol, from which Lincoln hugely benefited. Lincoln was in Springfield when a telegram arrived saying he had been nominated for president at the Republican National Convention in Chicago. He said: "I reckon there is a little short woman down in our house that would like to hear the news." He took his acceptance speech to the local school superintendent, who corrected a split infinitive.[45]

The Democratic papers dismissed Lincoln as "a third-rate lawyer," "a nullity," "a man in the habit of making coarse and clumsy jokes," one who "could not speak good grammar," a "gorilla." And we have to remember that most of Lincoln's sayings and speeches, and even his letters, have been cleaned up a good deal before coming down to us. The feeling that he was too tough to be president was not confined to the South, or even to Democrats. But William Cullen Bryant (1794–1878), the anti-slavery poet and philosopher, who had helped to found the Republican Party, called him "a poor flatboatman—such are *the true leaders of the nation.*" Lincoln had the Douglas Debates made into a little pamphlet, which he gave to people who asked his views. It served his purpose well. In dealing with the South's threat that his election would lead them to secede, he had already taken the bull by the horns in his speech at the Cooper Union in New York City, February 27, 1860: "You will not abide by the election of a Republican President! In that supposed event, you say, you will destroy the Union; and then, you say, the great crime of having destroyed the Union will be upon us! That is cool. A highwayman holds a pistol to my ear, and mutters through his teeth, 'Stand and deliver!—or I shall kill you, and then you will be a murderer!'"[46]

Using the political arithmetic of the previous thirty years, Lincoln should have been defeated. All the South had to do was to retain its links with the North, concentrate on keeping Jackson's old Democratic coalition together, and pick another Buchanan, or similar. But that was increasingly difficult to do, as the anti-slavers of the North raised the political temperature and the South replied with paranoia. Militant abolitionism dated from the early 1830s, when it became obvious that repatriating blacks to West Africa had failed—only 1,420 blacks had been settled in Liberia by 1831 and the number going there was declining. On January 1, 1831 William Lloyd Garrison (1805–79) began publishing the *Liberator* in Boston. It carried its motto on the front page: "I am in earnest—I will not equivocate—I will not excuse—I will not retreat a single inch—*and I will be heard.*" Garrison said he relied wholly on moral persuasion and condemned force, but some of his fiercest attacks were launched on moderate abolitionists and he began a new round of militancy on the Fourth of July 1854 when he burned a copy of the Constitution with the words, "So perish all compromises with tyranny." Meanwhile the American Anti-Slavery Society (1833) had been organized by two New York merchants, Arthur and Lewis Tappan, in conjunction with the most sophisticated and effective of the abolitionist campaigners, Theodore D. Weld (1803–95), whose anonymous tract, *American Slavery As It Is* (1839) furnished the inspiration for *Uncle Tom's Cabin.* Weld organized Oberlin as the first college to admit both blacks and women, and he married Angelina Grimke, one of two South Carolina sisters who freed their slaves and moved north to campaign.

Initially there was a lot of opposition to the anti-slavery movement in the North, where most Northerners hated blacks and frequently subjected them to mass violence. But by the end of the 1830s a younger generation who took the morality of abolition for granted began to take up positions and exercise influence. Emerson noticed "a certain tenderness in the people, not before remarked." As he put it, "The young men were born with knives in their brain." It was the beginning of liberal humanitarianism in the United States, and it took many forms, but slavery was the issue around which it concentrated. Increasingly, direct action of various kinds began to take over from propaganda alone. An underground railroad developed to get escaped slaves across the borders on to free soil and protect them there. It was run by "conductors" like Harriet Tubman (1821–1913), a Maryland slave who had escaped in 1849, the Quaker Levi Coffin (1789–1877), and the ferocious John Brown. There were about 1,000 conductors in all, and although their successes were numerically insignificant—not more than 1,000 a year after the passage of the Fugitive Slave Act of 1850, which made such operations increasingly risky—their effect on Southern morale was disproportionately great. Moreover, Southern slave-hunters, moving into Northern states in hot pursuit of fugitives, were highly unpopular especially when, as often happened, they grabbed the wrong black. From 1843 in Boston we get the first examples of an abolitionist mob releasing a recaptured fugitive slave by force. Whittier echoed the feelings of many with his lines:

No slave-hunt in our borders—no pirate on our strand!
No fetters in the Bay State—no slave upon our land!

During the 1850s, moreover, Northern legislatures passed laws making it exceedingly difficult, and sometimes impossible, to enforce the provisions of the 1850 federal act. The fact is that Southern aggression was all the time pushing Northern moderates into more extreme positions, particularly when the threat to the North's freedom of action became apparent. As William Jay, son of Chief Justice Jay, put it, "We commenced the present struggle to obtain the freedom of the slave—we are compelled to continue to preserve our own." James G. Birney (1792–1857), another former slave-owner who favored a modern position and was the Liberty Party candidate in 1840, put the point thus: "It has now become absolutely necessary that slavery should cease in order that freedom may be preserved in any portion of our land."[47]

As we have seen, from 1854 Kansas became the battleground of Southern extremists and anti-slavery activists. Indeed, it could be said that the Civil War started there. And it was inevitable, perhaps, that the kind of violence which became a daily occurrence in "bleeding Kansas" should spread. In particular, John Brown, who had received much applause for his "Pottawatomie Massacre"— "Brown of Pottawatomie" became a slogan of Northern militants—was given money and other help to set up a stronghold in the mountains of western Virginia to assist slaves traveling on the Underground Railroad. Not content with this, on October 16, 1859, with twenty men, he seized the U.S. arsenal at Harpers Ferry. Two days later, Colonel Robert E. Lee and a regular marine unit recaptured the post, killed ten of Brown's men, and made him prisoner. He was condemned to death and hanged on December 2.

Some, including Lincoln, condemned Brown; others, including Emerson, hailed him as "the new saint who will make the gallows glorious like the cross."[48] Brown's violent act completed the process of transforming the South, or at least its leadership class, into a tremulous and excitable body—a case of collective paranoia—which believed anything was preferable to a continuation of the present tension and fear. Some predicted a general rising of the slaves. Others looked to separation as the only safeguard of their property and way of life.

Against this background, the Democrats met for their presidential convention in April 1860 in Charleston, the South Carolina city which was the capital of Southern extremism. The Southerners, in their fear and fury, accused the Northern Democrats of betraying them by failing to present slavery to the North as a positive good. On behalf of the North, George E. Pugh of Ohio replied: "Gentlemen of the South, you mistake us—you mistake us—*we will not do it*." When the South failed to get the platform it wanted, the delegations from the Gulf states, South Carolina and Georgia, walked out, splitting the Democratic Party right down the middle. The convention met again at Baltimore on June 18 and finally nominated Douglas on a moderate platform. The Southerners replied by nominating the vice-president, John C. Breckinridge of Kentucky (1821–75), on a slavery platform. The Whigs reorganized themselves as the constitutional Union Party and nominated John Bell (1797–1869) of Tennessee as, in effect, the candidate of the border states. That meant four candidates. Essentially, however, it was a contest between Lincoln and Douglas in the North and Breckinridge and Bell in the South, since Lincoln could not hope to win Southern votes and Breckinridge had no support north of the Mason–Dixon Line.

In effect, a Lincoln victory was certain provided no untoward events intervened and provided he made no spectacular blunder. Hence all his friends and advisors warned him to keep out of the campaign and let the Republican Party do the work. So Lincoln worked behind the scenes to keep the Republican Party together, and left it to the Democrats, or rather the South, to commit political suicide. His only public appearance in the campaign was at Springfield in August where, pressed to orate, he simply said: "It has been my purpose, since I have been placed in my present position, to make no speeches." This gave him an almost Washingtonian detachment and saved him from misrepresentation. On November 6 Lincoln waited in the telegraph office until his victory in New York, signaled at 2 A.M. on the morning of the 7th, made his election certain. He got 1,866,452 votes against Douglas' 1,376,957; there had been 849,781 for Breckinridge and 588,879 for Bell. The result, in terms of electoral college votes, was somewhat different: Lincoln got 180, for he carried all but one of the free states, dividing New Jersey with Douglas (all the latter got, apart from Missouri). Breckinridge won all the slave states except Virginia, Tennessee, and Kentucky in the Upper South, which went to

Bell. In ten of the Southern states Lincoln did not receive a single vote. Moreover, he was elected on a minority vote of 39.9 percent, the lowest since J. Q. Adams won the unlucky, ominous election of 1824. The nation was indeed divided.[49]

If we now turn to Lincoln's principal opponent in the duel for the soul of America, we will see why it was that the South, having held so many cards in its hand, allowed itself to be exasperated into throwing away the game in a fit of temper. Jefferson Davis, Calhoun's political heir insofar as he had one, was president of the Confederacy from its reckless birth to its pitiful death-agony. He was flawed and blinkered both as man and as statesman, with huge weaknesses of judgment and capacity. But he was not small in any sense of the word. Six feet tall, slim, ramrod-straight, "soldierly bearing, a fine head and intellectual face . . . a look of culture and refinement about him," he "could infuse courage into the bosom of a coward, and self-respect and pride into the breasts of the most abandoned." To his cause he brought a passion "concentrated into a white heat, that threw out no sparks, no fitful flashes, glowing [instead] with an intense but not an angry glare." These judgments by contemporaries were endorsed even by critics and enemies. Thomas Cobb of Georgia said, "He is not great . . . [but] the power of will he has, made him all he is."[50]

The conventional portrait of Davis, the man driven by willpower, is of an old-fashioned Southern gentleman. That is inexact. His middle name was Finis because he was born when his mother was forty-seven, the last of ten. He had a modern-style upbringing: his father rejected any kind of corporal punishment, and the boy was cosseted by big sisters, and taught riding by his adoring big brothers, three of whom were old enough to have fought in the 1812 War. Jeff Davis was brought up to a simple, absolutist patriotism of a kind we would now find incomprehensible. When his father died, Davis' elder brother Joseph, a successful Mississippi cotton planter, took over the role of mentor and guardian. After an education under the Roman Catholic Dominican friars at Wilkinson County Academy and at the famous Transylvania University in Lexington, Kentucky, Davis went to West Point on the nomination of the war secretary, Calhoun, thereafter his political model and leader. As a frontier officer, he fought the Indians and personally took the surrender of Black Hawk, made peace among the miners and war against his superiors. Stiff-necked and bellicose, he admitted: "In my youth I was over-willing to fight." His career was checkered with rows, courts-martial, and frustration at slow promotion. When he married the daughter of General Zachary Taylor, he left the army and Brother Joseph set him up as a planter. This too was frustrating. Joseph owned 11,000 acres and was a wealthy man, but the 800-acre Hurricane Estate he "lent" or half-gave to Davis was small by Mississippi standards and he remained his brother's dependent.[51]

It is important to grasp that, when Davis spoke of the benevolence of the slave-system in the South, he believed what he said totally and spoke from experience. Joseph, as a planter, was enlightened. None of his slaves was ever flogged. The slaves judged and punished themselves. Families were kept together. One testified: "We had good grub and good clothes and nobody worked hard." Another: "Dem Davises never let nobody touch one of their niggers." The community at Davis Bend on the river, said General Taylor, was "a little paradise." Davis shared to the full his brother's attitudes and was anti-blood sports to boot. He treated his black body-servant, James Pemberton, with exquisite courtesy and put him in charge of his plantation when he was away. He made a point of returning any salute from a black with an elaborate bow: "I cannot allow any negro to outdo me in courtesy." Not for him the swaggering society of New Orleans or Charleston. His only genuflection to Southern male habits was a propensity to challenge critics to duels, though he never actually fought any. To sleep with one of his slaves would have been to him an abomination. When his beloved wife Sarah Taylor died of malaria, he acquired a sadness that never left him, though he eventually married again, a beautiful girl, Varina, half his age. His melancholy was aggravated by poor health, including terrifying facial pains and chronic hepatitis which eventually left him blind in one eye. He suffered from insomnia and his chief pleasure was reading—Virgil, Byron, Burns, and Scott.[52]

The overriding weakness of this seemingly civilized and well-meaning man was lack of imagination, compounded by ignorance. America in the 1840s and 1850s was already an immense country, but travel was still difficult, especially in the South, and expensive. It is hard for us to grasp how little Americans knew of the societies outside their region or indeed locality. Davis paid only one visit to New England and was surprised to find the people friendly. Until he became president of the Confederation he knew little of the South beyond his own part of Mississippi. He assumed that the treatment of slaves at Davis Bend was typical and refused to believe stories of cruelty: that was simply Northern malice and abolitionist invention. He was, like so many other well-read and well-meaning people in the South, the victim of its own policy of concentrating its limited media and publishing resources on indoctrinating its own people, and telling the rest of the world to go to hell. Davis was self-indoctrinated too; he had a passion for certitude.

On this narrowness of vision he built up a political philosophy which did not admit of argument. Blacks, he insisted, were better off as slaves in the South than as tribesmen in Africa: "I have no fear of insurrection, no more dread of our slaves than I have of our cattle . . . Our slaves are happy and contented." Not only was it in the interests of blacks to be slaves, it was likewise to their benefit that slavery be extended. Davis never possessed more than seventy-four slaves and knew all of them well: it was his policy. He maintained it was wrong for whites to own more slaves than they could personally care for, as he did. If cruelty occurred, it was because sheer numbers undermined the personal owner–slave relationship. So the more slavery spread out geographically, the more humane it would be. This was his argument for dismantling Mexico, turning its territories into new states, and making slavery lawful there and even north of the Missouri Compromise line. Slave-owners must be able to take their slaves with them into new territories just as immigrants had always taken any other form of property with them, such as wagons or cattle. Joseph had dinned into him the fundamental principle: "Any interference with the unqualified property of the owner in a slave was an abolition principle."

Davis believed that the Southern case for slavery and its extension rested on firm moral foundations. Indeed he was morally aggressive, accusing the North of hypocrisy: "You were the men who imported these negroes into this country. You enjoyed the benefits resulting from their carriage and sale; and you reaped the largest profits accruing from the introduction of the slaves." Abolition was nothing but "perfidious interference in the rights of other men." He did not see the agreements of 1820 and 1850 as "compromises" but as Southern concessions, the limit to which the South could reasonably be expected to go. Further limitations on slavery were merely Northern attacks on the South motivated not by morality but by envy and hatred: "The mask is off: the question is before us. It is a struggle for political power." The Constitution was on the South's side. The federal government had no natural authority: "It is the creature of the States.

As such it can have no inherent power; all it possesses was delegated by the States." If what Davis called "the self-sustaining majority" continued its oppressive and unlawful campaign against the South, the "Confederation" as he called it should be dissolved: "We should part peaceably and avoid staining the battlefields of the Revolution with the blood of a civil war."[53]

This philosophy, inherited from Calhoun and instilled by Brother Joseph, reexamined by Davis in his lonely musings, polished and consolidated over the years, he regarded as axiomatic. It is significant that he never saw himself as an extremist especially over breaking up the Union. He wrote: "I was slower and more reluctant than others. I was behind the general opinion of the people [of Mississippi] as to the propriety of prompt secession." But when his basic assumptions about slavery were challenged, he responded with paranoia. This sprang not just from his Southern conditioning but from a dominant streak of self-righteousness in his character. A variety of incidents in his early life, in the army, in his domestic and public quarrels show that, once he had made up his mind and adopted a position, he treated any attempt to argue him out of it as inadmissible, an assault on his integrity. As he put it to his second wife, Varina: "I cannot bear to be suspected or complained of, or misconstrued after explanation." That sentence sums up the tragedy of his life. Senator Isaac P. Walker of Wisconsin noted: "He speaks with an air which seems to say 'Nothing more can be said, I know it all, it must be as I think.'" Davis himself said he ignored press criticism: "Proud in the consciousness of my own rectitude, I have looked upon it with the indifference which belongs to the assurance that I am right."[54]

All this suggests that Davis was better suited to a military than a political life. That was Varina's view: "He did not know the arts of a politician, and would not practice them if understood." Davis got into politics in his later thirties but the Mexican War gave him the chance to resume his army career. He was elected colonel of a regiment of Mississippi volunteers, had the foresight to equip them with the new Whitney rifle, was favored by his commanding general and former father-in-law, General Taylor, saw action at Monterrey and Buena Vista, and distinguished himself in both these much publicized battles. The Mexican War, as we have noted, was the great proving ground for future American bigshots, both political and military. Davis was described by General Bliss, Taylor's chief-of-staff, as "the best volunteer officer in the Army," and President Polk offered him a general's commission. But he had been badly wounded in the foot at Buena Vista and chose instead to be nominated to the Senate.[55]

In politics Davis found it natural to be called the "Calhoun of Mississippi," and, when the old fire-eater died, to assume Elijah's Mantle. It was equally natural, when his friend Franklin Pierce became president, to accept office as war secretary, where he became perhaps the most powerful voice in the Cabinet and a forceful administrator.

But his weakness quickly made its appearance. He got into a series of arguments with his general-in-chief, Winfield Scott, mostly over trivialities. Scott was arrogant and self-righteous too, but Davis, as his political superior, might have been expected to behave with more sense and dignity. One of Davis' letters to Scott ran on for twenty-seven foolscap pages and was contemptuously described by its recipient as "a book." Everything fell into the hands of the press and made amazing reading. Scott closed his last letter: "Compassion is always due to an enraged imbecile," to which Davis replied that he was "gratified to be relieved of the necessity of further exposing your malignity and depravity."[56] Reading this correspondence helps to explain why the Civil War occurred and, still more, why it lasted so long. It certainly suggests that Davis was not a man fit to hold supreme office at any time, let alone during a war to decide the fate of a great nation.

It was not that Davis was unperceptive. In some ways his views were advanced. He tended to take the progressive line on everything except slavery. That pillar of Bostonian anti-slavery rectitude, John Quincy Adams, commended him warmly for helping to get the Smithsonian set up. And Davis was well aware of some of the South's weaknesses, especially its lack of industry. Its one big industrial complex was the Tredegar Iron Works on the banks of the James River near Richmond. It had been, as it were, replicated from the South Wales Tredegar works in the 1830s, to serve the Southern railroads. It also made cannon, chains, and iron ships, and by

1859 was the fourth-largest ironworks in the United States, employing 800 people. But it was near bankruptcy because it was uncompetitive. It got its iron ore from Pennsylvania because Virginian sources were exhausted, and virtually all its copper and bronze and many parts and machinery had to be bought in the North or from abroad. It had to pay extra wages because white industrial workers hated employment in a slave state. They particularly objected to working alongside slaves, fearing to be replaced by them. The works was notable for high labor turnover, chronic labor shortages, and neglect of innovation. It survived at all only because it gave liberal, risky credit to Southern railroads. It seemed enormous, and so reassuring, to Southerners, but in the nation as a whole it was marginal. There was in the South no central, up-to-date industrial magnet to attract skilled labor and so compensate for the many deterrents.[57]

By contrast, a hundred miles or so to the north there was the beginning of a vast manufacturing complex stretching from Wilmington to New York. From 1840 to 1860 this megalopolis was the most rapidly growing large industrial area in the world—and it was this complex which made inevitable, in military–economic terms, the South's ruin. Davis, knowing the South's weakness, began urging it, from about 1850 on, to start stockpiling arms and ammunition, to encourage immigration from the North, or to build railroads to transport its agricultural products itself, to create an industrial base to manufacture its own cotton goods, shoes,

hats, blankets, and so on, and to provide state support for higher education so that its sons were not
forced to go to Northern universities and adopt
their ideas. What finally happened to the South in
the 1950s, Davis was urging in the 1850s. But slavery repelled capital and white skilled labor alike,
and Southerners themselves did not want indus-
trialization for many different reasons, most of all
because they felt instinctively that it would mean
the end of slavery and plantation culture. So Davis
got no response to his pleas. In any case they were
half-hearted and confused. His wish to "educate"
the South conflicted with his insistence that Southern textbooks be rewritten to eliminate opinions in
conflict with the South's view of slavery, his desire
that the South's children should learn from books
which were "Politically Correct" and "indoctrinate
their minds with sound impressions and views" and
his determination to kick out "Yankee schoolteach-
ers." Not for nothing did the *New York Herald* call
him "the Mephistopheles of the South."[58]

By seceding from the Democratic Party, the Southern states threw away their greatest single asset, the
presidency. Then, by seceding from the Union, they
lost everything, slavery first and foremost. Bell was
right in proclaiming, throughout the election, that
the only way the South could retain slavery was by
staying in the Union. But that demanded "a change
of heart, radical and thorough, of Northern opinion
in relation to slavery."[59] Up to the beginning of the
campaign, Davis, realizing that Lincoln would win,

made a desperate effort to get all the other three candidates to withdraw in favor of a compromise figure—a sympathetic Northerner, perhaps. Breck-inridge and Bell agreed to stand down and so did Douglas' running mate, Benjamin Fitzpatrick. But Douglas, ambitious and self-centered—and blind—to the end, flatly refused. Thus Douglas made the Civil War inevitable. Or did he? *Was* it inevitable once Lincoln won?

One of the villains was Buchanan, the outgoing president, who in effect did nothing between the beginning of November 1860 and the handover to Lincoln in March. His message to Congress denied the right of secession but blamed the Republicans for the crisis—two incompatible opinions. He was lazy, frightened, confused, and pusillanimous. Thus four vital months were lost. His military dispositions, insofar as he made any, were inflammatory rather than conciliatory. Only two states wanted a civil war—South Carolina and Massachusetts. In the early 1830s over Nullification, the South Carolina extremists failed to carry anyone else with them, the rest of the South being prepared to trust President Jackson, to see the South got justice. But now they would trust nobody. All the same, an armed struggle might have been averted. Had South Carolina persuaded only four or five other states to go with it, the secession would have fizzled out. If all fifteen of the slave states had seceded, the North would have been forced to give way and sue for a compromise. As it was, just enough joined South Carolina to insure war.[60] The real tragedy for

America is that Lincoln, the man the South most hated, was exactly the man to get it to see reason, had he been given the chance. If he had been enabled by the Constitution to move into the White House immediately after his election, and assume full powers, all the weight of his intellect, and all the strength of his character, and all the genius of his imagination could have been brought to bear on the problem of exorcizing the South's fears. Instead, he had to sit, powerless (he used the interval to grow a beard), while the Union disintegrated, and by the time he took up command the process of secession was already taking place, and was irrevocable.[61]

As early as November 10, only three days after the election results were received, the South Carolina legislature unanimously authorized the election of a state convention on December 6, to decide "future relations between the State and the Union." Eight days later, Georgia followed suit. Within a month every state of the South had taken the initial steps towards secession. When Congress reassembled on December 3, it listened to a plaintive grumble from Buchanan, who said that he deplored talk of secession, but nothing could be done, by him anyway, to prevent it. Three days later South Carolina elected an overwhelmingly secessionist state convention which on December 20 declared that the state was no longer part of the Union. Davis himself tried to promote a compromise, then despaired of it. On January 7 the secession convention of his own state, Mississippi, met and on the 9th voted 84 to 15 to leave the Union. Two days before, the senators from Georgia, Florida, Alabama, Louisiana, Texas, Arkansas, and Mississippi had met in caucus in Washington and decided to meet again in Montgomery, Alabama on February 15 to form a government. Like other senators, Davis made an emotional speech of farewell in Congress. Going south through Tennessee, he was asked to make a speech at his hotel, Crutchfield House, and did so. Whereupon the brother of the hotel's owner, William Crutchfield, told him he was a "renegade and a traitor . . . We are not to be hoodwinked and bamboozled and dragged into your Southern, codfish, aristocratic, Tory-blooded South Carolina mobocracy." The crowd, many of them armed, backed these accusations—there was, strong Union sentiment in the back-country and in the mountains.[62]

Davis was promptly chosen general in Mississippi's army. Many, including his wife, wanted him to be commander-in-chief of the Confederate forces, rather than president. He agreed with Varina. Meeting on February 4, the six states which had already seceded, South Carolina, Mississippi, Florida, Georgia, Louisiana, and Alabama, drew up a new constitution, which was virtually the same as the old except it explicitly recognized slaves as property. Robert Toombs (1810–85), Senator from Georgia, might have got the presidency, but he got publicly drunk several nights running. In the end Davis was chosen more or less unanimously. His journey from his home near Vicksburg to his inauguration in Montgomery was a sinister foretaste of the problems the South faced. The two cities were less than 300 miles apart, along a direct east-west road, but Davis, trying to get there more quickly by rail, had to travel north into Tennessee, then across northern Alabama to Chattanooga, south to Atlanta, and from there southwest to Montgomery, a distance of 850 miles around three-and-one-half sides of a square on half a dozen different railroads using three different gauges. No railway trunk lines bound the rebellious states together. The South had no infrastructure.[63] Its railroad system was designed solely to get cotton to sea for export. There was virtually no interstate trade in the South, and so no lines to carry it. It took five railroad lines to get from Columbia to Milledgeville, for example; the railroads in Florida, Texas, and most of Louisiana had no connection at all with the other Southern states. The functional geography of the South, both natural and manmade, was against secession.

In his inaugural, Davis said the Confederacy was born of "a peaceful appeal to the ballot box." That was not true. No state held a referendum. It was decided by a total of 854 men in various secession conventions, all of them selected by legislatures, not by the voters. Of these 157 voted against secession. So 697 men, mostly wealthy, decided the destiny of 9 million people, mostly poor. Davis said he was anxious to show that secession was "not a rich man's war and a poor man's fight," but the fact is it was the really rich, and the merely well-to-do, both of whom had a major interest in the struggle, who decided to commence it, not the rest of the whites, who had no direct economic interest at all. And the quality of Southern leadership, intellectually at least, was poor. The reasons for secession, put into the declarations of each states, made no sense, and merely reflected the region's paranoia. Mississippi's said: "the people of the Northern states have assumed a revolutionary position towards the Southern states." They had "insulted and outraged our citizens when traveling amongst them . . . by taking their servants and liberating the same." They had "encouraged a hostile invasion of a Southern state to incite insurrection, murder and rapine." South Carolina's was equally odd, ending in a denunciation of Lincoln, "whose opinions and purposes are hostile to slavery." But most presidents of the United States had been hostile to slavery, not least Jefferson, the man whose opinions on the subject Lincoln most often quoted.

The Southern leaders assumed there were absolute differences between the peoples of North and South. In fact allegiances were divided. Mary Lincoln had three brothers in the Confederate Army, all of whom were killed—and her emotional sympathies were certainly with the South. Varina Davis' male relatives, the Howells, were all in the Union Army. Senator John J. Crittenden of Kentucky (1787–1863), who did his best to promote compromise, had two sons, both major-generals, one serving in the Confederate, the other in the Union army. The best Union agent in Europe, Robert J. Walker, was a former senator from Mississippi, while the best Confederate agent, Caleb House, came from Massachusetts. General Robert E. Lee's nephew,

Samuel P. Lee, commanded the Union naval forces on the James River, while another Union admiral, David Glasgow Farragut (1801–70), the outstanding maritime commander in the war, was born in Tennessee and lived in Virginia. The examples are endless. The young Theodore Roosevelt was made to pray for the North; the young Woodrow Wilson prayed for the South. There were, literally, millions of divided families, and the number of extremists on both sides probably did not amount to a hundred thousand all told.

It became a necessity, Jefferson Davis wrote to a Northern friend, January 20, 1861, "to transfer our domestic institutions from hostile to friendly hands, and we have acted accordingly." Lincoln could not exactly be called friendly towards the South—he was, rather, exasperated and sad. But he was not hostile. Southern leaders like Davis would not accept that Lincoln was hated by many abolitionists, like Wendell Phillips (1811–84), the rich Boston humanitarian ideologue, who called him "the Slavehound of Illinois." The most the Lincoln Republicans could do, and proposed to do, was to contain slavery. To abolish it in the 1860s required a constitutional amendment, and a three-quarters majority; as there were fifteen slave states, this was unobtainable. A blocking majority of this magnitude would still have been sufficient in the second half of the 20th century. It is worth noting that, at the time of secession, Southerners and Democrats possessed a majority in both houses of Congress, valid till 1863 at least. If protecting slavery was the aim, secession made no sense. It made the Fugitive Slave Act a dead letter and handed the territories over to the Northerners. The central paradox of the Civil War was that it provided the only circumstances in which the slaves could be freed and slavery abolished.[64]

War was so obviously against the rational interests of the South that Lincoln did not consider it likely. His concern was to prevent the Republicans from appeasing the South by abandoning their platform and embracing Douglas' popular-sovereignty doctrine. Over and over again he repeated his message to Republican congressmen: "Have none of it. Let there be no compromise on the question of *extending* sovereignty. Stand firm. The tug has to come and better now, than any time hereafter."[65] By tug, he meant confrontation and crisis, not war. If he had thought in terms of war when appointing his Cabinet, Lincoln would never have made Simon Cameron (1799–1889) his secretary of war. Cameron was a millionaire banker and railroad tycoon, who was the overwhelming boss of Pennsylvanian Republicanism and he was appointed for entirely political reasons (his handling of army contracts led Lincoln to sack him and to a vote of censure in the House). Nor, probably, would he have made Seward secretary of state and Chase treasury secretary. Lincoln knew a vertiginous time was ahead and he opted for a strong government rather than a warlike one.[66]

Seward, a clever, persuasive man, believed the administration's best strategy was to leave the rebellious Deep South to stew in its own Confederate juice and concentrate on wooing the other slave states to remain faithful to the Union. But that would have meant letting the seven go, and Lincoln was determined to preserve the Union as it was, at all costs. That was the only thing which, at this stage, he could see clear, and he stuck to it. This strategy, in turn, set off the mechanism of the war. On asserting its independence in December 1860, South Carolina called on the custodians of all federal property within the state to surrender it. Major Robert Anderson, the federal commander of the fortifications in Charleston Harbor, concentrated his forces in Fort Sumter and refused to act without instructions from Washington; President Buchanan, lax in most ways, likewise declined to have the federal forces evacuated. General P.G.T. Beauregard of South Carolina thereupon trained his guns on the Fort. When Lincoln took over, the Cabinet deliberated on what to do. The commander-in-chief, General Scott, who might have been expected to be anxious to poke his old "imbecile" enemy Davis in the eye, in fact advised doing nothing. Five out of seven members of Lincoln's Cabinet agreed with him. But Lincoln decided otherwise. His decision to send a relieving expedition by sea, carrying food but no arms or ammunition, and to inform South Carolina of what he was doing, demonstrated his policy of upholding the Union at any cost. The response of the Confederate forces, to fire on the Fort, and the flag, was a decision to secede at any cost. That began the war on April 12, 1861.[67]

As the South was arming and recruiting, Lincoln had no alternative but to take steps too. "The star-spangled banner has been shot down by Southern troops," he said, and on April 15 asked for 75,000 volunteers (answered by 92,000 within days). This move by Lincoln was, curiously enough, the "last straw" which pushed Virginia (and so North Carolina) into secession. This, too, was undemocratic since the state convention voted 88 to 55, on April 17, to submit an Ordinance of Secession to a popular plebiscite. However, the governor put the state under Confederate command without waiting for the vote. This event was decisive for many reasons. Virginia was the most important of the original colonies, the central element in the Revolutionary War, and the provider of most of the great early presidents, as well as of the U.S. Constitution itself. For the state which had done more than any other to bring the Union into existence to leave it in such an underhand and unconstitutional manner was shabby beyond belief. It is astonishing that the Virginians put up with it. And of course many of them did not. The people of West Virginia, who had no slaves, broke off and formed a separate state of their own, acknowledged by Congress as the State of West Virginia in 1863.

General Lee, the state's most distinguished soldier, had been asked by Lincoln to become commander-in-chief of the Union forces. This was a wise choice and would have been a splendid appointment, for Lee was decent, honorable, and sensible as well as skillful. But Lee was a Virginian before anything else and he waited to see what Virginia did. When Virginia seceded, he reluctantly resigned his commission in the U.S. Army, which he had served for thirty-two years. It seems to us quixotic but he felt he had no other option. He wrote to his sister in Baltimore and his brother in Washington, D.C.: "With all my devotion to the Union, and the feeling of loyalty and duty of an American citizen, I have not been able to make up my mind to raise my hand against my relatives, my children, my home."[68]

Arkansas seceded on May 6. The next day Tennessee formed an "alliance" with the Confederacy, the only decision to be endorsed by popular vote. North Carolina, sandwiched between Virginia and South Carolina, had not much choice and joined on May 20. Missouri was divided but refused to join the Confederacy. Delaware was solid for the Union but shaky on coercion. Maryland too protested against coercion but declined to summon a state convention and so remained in the Union. Kentucky initially refused to send volunteers at Lincoln's request but by the end of 1861 had joined the Union war-effort. So only eleven out of the fifteen slave states formed the Confederacy.[69]

In demographic terms, the Confederacy was at a huge disadvantage. The census of 1860 showed that the eleven Confederate states had a population of 5,449,467 whites and 3,521,111 slaves. Nearly 1 million of the white males served, of whom 300,000 were casualties. The nineteen Union states had a population of 18,936,579 and the four border states a further 2,589,533, plus 429,401 then-slaves, over 100,000 of whom served in the Union Army, which altogether numbered 1,600,000. Moreover, during the war nearly a million further immigrants arrived in the North, of whom 400,000 served in the Union Army. Some of the best Northern troops were German, Irish, and Scandinavian, as were some of the smartest officers—Franz Sigel, Carl Schurz (Germans), Philippe de Trobriand (French), Colonel Hans Christian Heg and Hans Matson (Norwegian), and Generals Corcoran and Meagher (Irish). The Union economic preponderance was even more overwhelming. If the North–South ratio in free males aged eighteen to sixty was 4.4:1, it was 10:1 in factory production, iron 15:1, coal 38:1, firearms production 32:1, wheat 412:1, corn 2:1, textiles 14:1, merchantship tonnage 25:1, wealth 3:1, railroad mileage 2.4:1, farm acreage 3:1, draft animals 1.8:1, livestock 1.5:1. The only commodity in which the South was ahead was cotton, 24:1, but this advantage was thrown away by overproduction (in the South) and stockpiling (outside the South) in the endless build-up to the crisis. Just before the war, Senator James Henry Hammond of South Carolina boasted: "Cotton, rice, tobacco and naval stores command the world; and we have the sense to know it, and are sufficiently Teutonic to carry it out successfully. The North without us would be a motherless calf, bleating about, and die of mange and starvation."[70] The assumption in the South was that the coming of war would lead to an expansion of its economy, and a contraction of the North's. In fact, as was foreseeable, the reverse occurred. The South's economy shrank, the North's expanded, even faster than in the 1850s.[71]

The South compounded its difficulties by weaknesses in its handling of finance, diplomacy, and internal politics, all of which had severe military consequences. First, it is a curious historical fact that most civil wars are lost by one side running out of money, and the American Civil War was an outstanding case in point. The South had no indigenous gold or silver supplies and no bullion reserves, and was entirely dependent on its own paper money. The North had the enormous advantage of a large, well-trained navy and, almost from the start, was able to impose a blockade, often ineffective at first but progressively tighter as the war proceeded. As a result, import and export taxes, the way of raising money traditionally preferred by the South, raised little. Import duties brought in only about $1 million in specie during the entire war, and the Union navy was so vigilant in running down cotton-export ships that only about $6,000 in specie was collected from cotton exports. With its limited capacity to produce armaments, the South was forced to shop abroad. France, always happy to supply arms to dodgy-regimes, duly obliged but insisted on being paid in specie (as did independent gun-runners).

As his Treasury secretary, Davis appointed C. G. Memminger, a local South Carolina politician. This was an extraordinary choice: Memminger had virtually no experience of finance and, more important, lacked the creative ingenuity to surmount the almost insuperable difficulties of raising hard cash.[72] An initial war-loan of 8 percent, organized by a consortium of New Orleans and Charleston banks, raised $15 million in specie, all of which was immediately sent abroad to buy arms. But subsequent loans were relative, then total, failures. A cotton-backed foreign loan, organized in London by Erlangers in January 1863, brought in disappointingly little, as a result of high charges and an imprudent attempt to bull the market. Hence Memminger resorted to the device of the improvident through the ages—printing paper. By the summer of 1861, $1 million of Confederate paper currency was circulating. By December it was over $30 million; by March 1862 $100 million; August 1862 $200 million; December 1862 $450 million. In 1863 it doubled again to $900 million and continued to increase, though later figures are mere guesswork. Gold was quoted at a premium over paper as early as May 1861 and was 20 percent premium by the end of the year. By the end of 1862 a gold dollar bought three paper ones and, by the end of 1863, no fewer than twenty.

In July 1864 Memminger, accused of making private profits on cotton-running, resigned in disgust, and Davis then appointed a real economic wizard called George A. Trenholm, a Charleston cotton-merchant who had proved extraordinarily adept at selling the South's staple. But by then it was too late: the South's finances were beyond repair. Inflation became runaway, the gold dollar being quoted at 40 paper ones in December 1864 and 100 shortly thereafter. Inflation, if nothing else, doomed the South. In the second half of the war Southerners showed an increasing tendency to use the North's money, as it inspired more confidence. Towards the end people cut themselves off from paper money altogether, and bought and sold in kind—even the government raised taxes and loans in produce. The only people with means to move around were those who had kept gold dollars. Davis was like everyone else. In the final weeks of the Confederacy he sent his wife Varina off with his last remaining pieces of gold, keeping one five-dollar coin for himself.[73]

The South's diplomacy was as inept as its finance. Davis did not initially see the need for a major diplomatic effort since he believed the economic arguments would speak for themselves. The key country was Britain, because in the 1850s it had imported 80 percent of its cotton from America, and it had the world's largest navy, which could break the Union blockade if it wished. Davis accepted Senator Hammond's assertion: "You dare not make war upon our cotton. No power on earth dares make war on it. Cotton is King."[74] But overproduction and stockpiling in anticipation of war led to a 40 percent oversupply of cotton in the British market by April 1861, before the war had properly begun. Britain got cotton from Egypt and India and, later in the war, from the United States itself, via the North. In the years 1860–65 Britain managed to import over 5 million bales of cotton from America, little of which was bought from the South directly. British manufacturers welcomed the opportunity to work off stocks and free themselves from dependence on Southern producers, whom they found difficult and arrogant. It is true the cotton blockade caused some unemployment in Lancashire and Yorkshire—by the end of 1862 it was calculated that 330,000 men and women were out of work in Britain as a result of the conflict. But they had no sympathy for the South. They identified with the slaves. They sent a petition to Lincoln: "Our interests are identical with yours. We are truly one people . . . If you have any ill-wishers here, be assured they are chiefly those who opposed liberty at home, and that they will be powerless to stir up quarrels between us." Lincoln called their words "an instance of sublime Christian heroism."

The truth is, by opposing slavery and by insisting on the integrity of the Union, Lincoln identified himself and his cause with the two most powerful impulses of the entire 19th century—liberalism and nationalism. He did not have to work at a powerful diplomatic effort—though he did—as world opinion was already on his side, doubly so after he issued his Emancipation Proclamation. It was the South which needed to put an effort into winning friends. It was not forthcoming. Davis hated Britain anyway. The South had many potential friends there—the Conservative Party, especially its leading families, newspapers like *The Times,* indeed a surprisingly large section of the press. But he did not build on this. The envoys he sent were extremists, who bellowed propaganda rather than insinuated diplomacy. The British Prime Minister, Lord Palmerston, was a Whig–Liberal nationalist who played it cool: on May 13, 1861, he declared "strict and impartial neutrality." The North's naval blockade caused much less friction with Britain than the South had hoped, because it conformed strictly to British principles of blockading warfare, which the Royal Navy was anxious to see upheld for future use. The one really serious incident occurred in November 1861, when the famous explorer Captain Charles Wilkes (1798–1877), commanding the USS *San Jacinto*, stopped the British steamer *Trent* and seized two Confederate commissioners, John Slidell and James M. Mason. This caused an uproar in Britain, but Seward, as secretary of state, quickly defused the crisis by ordering the men's release, on the ground that Wilkes should have brought the ship into harbor for arbitration.[75]

Added to improvident economics and incompetent diplomacy, the South saddled itself with a political system which did not work. It was a martyr to its own ideology of states' rights. Although Davis and his fellow-Southerners were always quoting history, they did not know it. Had they studied the early history of the republic objectively, they would have grasped the point that the Founding Fathers, in drawing up the Constitution, had to insure a large federal element simply because the original provisional system did not work well, in war or in peace. The Confederacy thus went on to repeat many of the mistakes of the early republic. Each state raised its own forces, and decided when and where they were to be used and who commanded them. To many of their leaders, the rights of their state were more important than the Confederacy itself. Men from one state would not serve under a general from another. Senior commanders with troops from various states had to negotiate with state governments to get more men. Davis had to contend with many of the identical difficulties, over men and supplies and money, which almost overwhelmed Washington himself in the 1770s—and he had none of Washington's tact, solidity, resourcefulness, and moral authority. Everyone blamed him, increasing his paranoia. As a former military man and war secretary, he thought he knew it all and tried to do everything himself. When he set up his office, he had only one secretary. His first secretary of war, Leroy P. Walker, was a cipher. Visitors noticed Davis summoned him by ringing a desk bell, and Walker then trotted in "exhibiting a docility that dared not say 'nay' to any statement made by his chief."[76] Congress refused to take account of any of his difficulties and behaved irresponsibly—it was composed mainly of vainglorious extremists. Davis had more trouble with his congress than any Union president, except possibly Tyler. He vetoed thirty-eight bills and all but one later passed with Congress overriding his veto. Lincoln had to use the veto only three times, and in each case it stuck.

But many of Davis' difficulties were of his own making. His constant illnesses did not help, as during them he became short-tempered and dictatorial. As his absurd row with Scott showed, he could not distinguish between what mattered and what was insignificant. Virtually all his early appointments, both Cabinet and army, proved bad. Davis resumed personal vendettas going back to the Mexican War and even to his West Point days. In the South, everyone knew each other and most had grudges. In picking senior commanders, Davis favored former West Point classmates, war-service comrades, and personal friends. Things were made even more difficult by each state demanding its quota of generals, and by muddles Davis made over army regulations. A lot of his bitterest rows with colleagues and subordinates had nothing to do with the actual conduct of the war. The Navy Secretary Stephen Mallory (1813–73), a Trinidadian and one of the few Confederate leaders who knew what he was doing, deplored the fact that "our fate is in the hands of such self-sufficient, vain, army idiots." Davis was not the man to run difficult generals, and he became almost insensate with rage when he was personally blamed for lack of men and supplies, above all lack of success. Varina admitted: "He was abnormally sensitive to disapprobation. Even a child's disapproval discomposed him . . . and the sense of mortification and injustice gave him a repellent manner." Faced with criticism he could not bear, he took refuge in illness.[77]

A lot of Davis' strategic difficulties were his own fault. Despite conscripting 90 percent of its able white manpower, the South was always short of troops. In January 1862 its army rolls numbered 351,418, against a Union strength of 575,917. It reached its maximum in January 1864, when 481,180 were counted under the Confederate flag.

Therafter the South's army declined in strength
whereas the North's rose, so that in January 1865
the respective numbers were 445,203 and 959,460.[78]
That being so, Davis should have concentrated his
smaller forces in limited areas. Instead, he took
seriously and followed to the letter his inauguration oath to defend every inch of Confederate territory. This was an impossible task. It involved, to
begin with, defending over 3,500 miles of coastline,
without a navy to speak of. Texas alone had 1,200
miles of border. If Kentucky had seceded, it would
have provided a simple water-border. For a time it
kept out both sides, but eventually the Unionists
menaced the South from there too. Missouri was
also divided but its settled eastern reaches, centered
on St. Louis, were firmly Unionist, and that left an
almost indefensible 300-mile straight-line border in
northern Arkansas. Hence a large percentage of the
Confederate army, perhaps a third or even more,
was always employed on non-combative defensive
duties when its active commanders were clamoring
desperately for troops. It is true that the Unionists
also used vast numbers of men on the gradually
extending lines of communication—but then they
had more men to use.[79]

Early in the war the Confederate capital was
moved from Montgomery to Richmond, mainly to
ensure that Virginia stayed committed to the fight.
This was a mixed blessing. The polished Virginians regarded the South Carolinans, who formed
the core of the government, as loudmouthed, flashy,
dangerous extremists. They looked down their

noses at the Davises. The ladies noted Varina's dark color and thick lips, comparing her to "a refined mulatto cook" and called her the "Empress," a reference to the much-despised Eugénie, wife of the French dictator, Napoleon III. The Georgians, especially Thomas Cobb, were hostile to Davis: he was, said Cobb, as "obstinate as a mule," and they dismissed J. P. Benjamin (1811–84), the attorney-general and by far the ablest member of the Confederate government, as a "Jew dog." Senator Louis T. Wigfall of Texas was a strong Davis supporter until their wives fell out, wherupon Charlotte Wigfall, a South Carolina snob, called Varina "a coarse, western woman" with "objectionable" manners, and Wigfall preached mutiny and sedition in the Congress, often when drunk. Confederate Richmond gradually became a snakepit of bitter social and political feuds, and the Davises ceased to entertain.

Once Northern armies began to penetrate Confederate soil, the interests of the states diverged and it was everyone for himself, reflected in Richmond's savage political feuding. It is a curious paradox that ordinary Southerners, who had not been consulted, fought the war with extraordinary courage and endurance, while their elites, who had plunged them into Armageddon, were riven by rancorous factions and disloyalty, and many left the stricken scene long before the end.[80] Davis was too proud, aloof, and touchy to build up his own faction. He thought it beneath him to seek popularity or to flatter men into doing their duty. Hence "close friends sometimes left shaking their heads or fists, red with

anger and determined never to call on him again."[81]
But at least he went down with the stricken cause,
ending up in Unionist fetters.

It may be asked: all this being so, why did the South
fight so well? Why did the war last so long? In the
first place, it has to be understood that Lincoln was
operating under many restraints. He did not seek
war, want war, or, to begin with, consider he was
in any way gifted to wage it. He made a lot of mistakes, especially with his generals, but unlike Davis
he learned from them. The South was fighting for
its very existence, and knew it; there was never any
lack of motivation there. The North was divided,
bemused, reluctant to go to war; or, rather, composed of large numbers of fanatical anti-slavers
and much larger numbers of unengaged or indifferent voters who had no wish to become involved
in a bloody dispute about a problem, slavery, which
did not affect them directly. Then there were the
four border states, all of them slave-owning, whose
adherence to the Union it was essential to retain.
Lincoln, beginning with a professional army of a
mere 15,000, was fighting a war waged essentially
for a moral cause, and he had to retain the high
moral ground. But he had also to keep the rump
of the Union together. That meant he had to be
a pragmatist without ever descending into oppor-
tunism. His great gift—perhaps the greatest of
the many he possessed—was precisely his ability
to invest his decisions and arguments with moral
seemliness even when they were the product of

empirical necessity. He was asked to liberate the slaves—what else was the war about? He answered: it was to preserve the Union. He realized, he knew for a fact, that if he did preserve the Union, slavery would go anyway. But he could not exactly say so, since four of his states wanted to retain it.

Some of Lincoln's generals, for military purposes, began to issue local emancipation decrees, hoping to get the Southern slaves to rise and cause trouble behind Confederate lines. Lincoln had to disavow these efforts as *ultra vires*. He hated slavery. But he loved the Constitution more, writing to a friend in Kentucky:

> I am naturally anti-slavery. If slavery is not wrong, nothing is wrong. I cannot remember when I did not so think and feel, and yet I have never understood that the presidency conferred on me an unrestricted right to act officially on this judgment and feeling. It was in the oath I took that I would, to the best of my ability, preserve, protect and defend the Constitution of the United States. I could not take the office without taking the oath. Nor was it my view that I might take an oath to get power, and break the oath in using the power.

He made public his intentions about slavery in an order disavowing an emancipation decree issued by General David Hunter. Declaring it "altogether void" and rejecting the right of anyone except himself to liberate the slaves, he nonetheless made it publicly clear that such a right might well be invested in his presidential power: "I further make it known that whether it be competent for me, as Commander-in-Chief of the Army and Navy, to declare the slaves of any State or States free, and whether at any time and in any case, it shall have become a necessity indispensable to the maintenance of the Government to exercise such supposed power, are questions which, under my responsibility, I reserve to myself, and which I cannot feel justified in leaving to the decision of commanders in the field."[82]

He followed this up by writing a reply to Horace Greeley, who had published a ferocious editorial in the *New York Tribune,* entitled "The Prayer of Twenty Millions," accusing Lincoln of being "strangely and disastrously remiss" in not emancipating the slaves, adding that it was "preposterous and futile" to try to put down the rebellion without eradicating slavery.[83] Lincoln replied by return of post, without hesitation or consultation, and for all to read:

My paramount object in this struggle is to save the Union and it is not either to save or to destroy slavery. If I could save the Union without freeing any slaves, I would do it; and if I could save it by freeing all the slaves I would do it; and if I could save it by freeing some slaves and leaving others alone I would do that. What I do about slavery and the colored race I do because I believe it helps to save the Union . . . I shall do less whenever I believe that what I am doing hurts the cause, and I shall do more whenever I believe doing more helps the cause.[84]

In seeking to keep the Union together, and at the same time do what was right by the slaves, the innocent victims as well as the cause of the huge convulsive struggle, Lincoln was fully aware that the Civil War was not merely, as he would argue, an essentially constitutional contest with religious overtones but also a religious struggle with constitutional overtones. The enthusiasts on both sides were empowered by primarily moral and religious motives, rather than economic and political ones. In the South, there were standard and much quoted texts on negro inferiority, patriarchal and Mosaic acceptance of servitude, and of course St Paul on obedience to masters. In the events which led up to the war, both North and South hurled texts at each other. Revivalism and the evangelical movement generally played into the hands of extremists on both sides.[85] When the war actually came, the Presbyterians, from North and South, tried to hold together by suppressing all discussion of the issue; but they split in the end. The Congregrationalists, because of their atomized structure, remained theoretically united but in fact were divided in exactly the same way as the others. Only the Lutherans, the Episcopalians, and the Catholics successfully avoided public debates and voting splits; but the evidence shows that they too were fundamentally divided on a basic issue of Christian principle.[86]

Moreover, having split, the Christian churches promptly went to battle on both sides. Leonidas Polk, Bishop of Louisiana, entered the Confederate army as a major-general and announced: "It is for constitutional liberty, which seems to have fled to us for refuge, for our hearth-stones and our altars that we fight." Thomas March, Bishop of Rhode Island, preached to the militia on the other side: "It is a holy and righteous cause in which you enlist . . . God is with us . . . the Lord of Hosts is on our side." The Southern Presbyterian Church resolved in 1864: "We hesitate not to affirm that it is the peculiar mission of the Southern Church to conserve the institution of slavery, and to make it a blessing both to master and slave." It insisted that it was "unscriptural and fanatical" to accept the dogma that slavery was inherently sinful: it was "one of the most pernicious heresies of modern times."

To judge by the hundreds of sermons and specially composed church prayers which have survived on both sides, ministers were among the most fanatical of the combatants from beginning to end. The churches played a major role in dividing the nation, and it may be that the splits in the churches made a final split in the nation possible. In the North, such a charge was often willingly accepted. Granville Moddy, a Northern Methodist, boasted in 1861: "We are charged with having brought about the present contest. I believe it is true we did bring it about, and I glory in it, for it is a wreath of glory round our brow." Southern clergymen did not make the same boast but of all the various elements in the South they did the most to make a secessionist state of mind possible. Southern clergymen were particularly responsible for prolonging the increasingly futile struggle. Both sides claimed vast numbers of "conversions" among their troops and a tremendous increase in churchgoing and "prayerfulness" as a result of the fighting.[87]

The clerical interpretation of the war's progress was equally dogmatic and contradictory. The Southern Presbyterian theologian Robert Lewis Dabney blamed what he called the "calculated malice" of the Northern Presbyterians and called on God for "a retributive providence" which would demolish the North. Henry Ward Beecher, one of the most ferocious of the Northern clerical drumbeaters, predicted that the Southern leaders would be "whirled aloft and plunged downward forever and ever in an endless retribution." The New Haven theologian Theodore Thornton Munger declared, during the "March through Georgia," that the Confederacy had been "in league with Hell," and the South was now "suffering for its sins" as a matter of "divine logic." He also worked out that General McClellan's much criticized vacillations were an example of God's masterful cunning since they made a quick Northern victory impossible and so insured that the South would be much more heavily punished in the end.[88]

As against all these raucous certainties, there were the doubts, the puzzlings, and the agonizing efforts of Abraham Lincoln to rationalize God's purposes. To anyone who reads his letters and speeches, and the records of his private conversations, it is hard not to believe that, whatever his religious state of mind before the war again, he acquired faith of a kind before it ended. His evident and total sincerity shines through all his words as the war took its terrible toll. He certainly felt the spirit of guidance. "I am satisfied," he wrote, "that when the Almighty wants me to do or not to do a particular thing, he finds a way of letting me know it." He thus waited, as the Cabinet papers show, for providential guidance at certain critical points of the war. He never claimed to be the personal agent of God's will, as everybody else seemed to be doing. But he wrote: "If it were not for my firm belief in an overriding providence it would be difficult for me, in the midst of such complications of affairs, to keep my reason in its seat. But I am confident that the Almighty has his plans and will work them out; and . . . they will be the wisest and the best for us." When asked if God was on the side of the North, he replied: "I am not at all concerned about that, for I know the Lord is always on the side of the right. But it is my constant anxiety and prayer that I and this nation should be on the Lord's side." As he put it, "I am not bound to win but I am bound to be true. I am not bound to succeed, but I am bound to live up to the light I have."[89]

Early in the war, a delegation of Baltimore blacks presented him with a finely bound Bible, in appreciation of his work for the negroes. He took to reading it more and more as the war proceeded, especially the Prophets and the Psalms. An old friend, Joshua Speed, found him reading it and said: "I am glad to see you so profitably engaged." Lincoln: "Yes. I *am* profitably engaged." Speed:

"Well, I see you have recovered from your skepticism [about religion and the progress of the war]. I am sorry to say that I have not." Lincoln: "You are wrong, Speed. Take all of this book upon reason that you can, and the balance on faith, and you will live and die a happier and a better man." As he told the Baltimore blacks: "This Great Book . . . is the best gift God gave to man."[90] After reading the Bible, Lincoln argued within himself as to what was the best course to pursue, often calling in an old friend like Leonard Swett, to rehearse pros and cons before a sympathetic listener.

Thus arguing within himself, Lincoln incarnated the national, republican, and democratic morality which the American religious experience had brought into existence—probably more completely and accurately than a man committed to a specific church. He caught exactly the same mood as President Washington in his Farewell Message to Congress, and that is one reason why his conduct in the events leading up to the war, and during the war itself, seems, in retrospect—and seemed so to many at the time—so unerringly to accord with the national spirit. Unlike Governor Winthrop and the first colonists, Lincoln did not see the republic as the Elect Nation because that implied it was always right, and the fact that the Civil War had occurred at all indicated that America was fallible. But, if fallible, it was also anxious to do right. The Americans, as he put it, were "the Almost Chosen People" and the war was part of God's scheme, a great testing of the nation by an ordeal of blood, showing the way to charity and thus to rebirth.

In this spirit Lincoln approached the problem of emancipating the slaves. The moment had to be well chosen not merely to keep the border states in the war, and fighting, but because in a sense it marked a change in the object for which the war was being fought. Lincoln had entered it, as he said repeatedly, to preserve the Union. But by the early summer of 1862 he was convinced that, by divine providence, the Union was safe, and it was his duty to change the object of the war: to wash away the sin of the Constitution and the Founding Fathers, and make all the people of the United States, black as well as white, free. Providence had guided him to this point; now providence would guide him further and suggest the precise time when the announcement should be made, so as to bring victory nearer.

Lincoln had weighed all the practical arguments on either side some time before he became convinced, for reasons which had little to do with political factors, that the slaves should be declared free, and laid his decision before the Cabinet on July 22. He told his colleagues he had resolved upon this step, and had not called them together to ask their advice but "to lay the subject-matter of a proclamation before them." Their response was pragmatic. Edwin M. Stanton (1814–69), secretary of war, and Edward Bates (1793–1869), attorney-general, urged "immediate promulgation" for maximum effect. Chase thought it would unsettle the government's financial position. Postmaster-General

Montgomery Blair (1813–83) said it would cost them the fall elections. Lincoln was unperturbed. The decision was taken: all that was now required was guidance over the timing. "We *mustn't issue it* until after a victory," he said, many times. That victory came, as he knew it would, on September 17, with Antietam. Five days later, on September 22, the Emancipation Proclamation, the most revolutionary document in United States history since the Declaration of Independence, was made public, effective from January 1, 1863. Despite an initially mixed reception, the ultimate impact of this move on the progress of the war was entirely favorable—as Lincoln, listening to the heedings of providence, knew it would be.[91]

Political considerations—holding the Union together, putting his case before world opinion, in which emancipation played a key part, satisfying his own mind that the war was just and being justly pursued—were not the only considerations for Lincoln, or even the chief ones. The overriding necessity, once the fighting began, was to win, and that Lincoln found the most difficult of all. His problem was not providing the men and the supplies, or the money to pay for them. The money was spent on a prodigious scale, and soon exceeded $2 million a day. At the outset of the conflict, the U.S. public debt, which had risen slowly since President Jackson wiped it out, was a little under $70 million. By January 1, 1866, when the end of the insurrection was officially proclaimed, it stood at $2,773 million. But Congress was willing to vote heavy taxes including, for the first time, a tax on personal incomes of from 3 to 5 percent (it was phased out in 1872). All the same, payments in specie had to be suspended at the end of December 1861, and in February 1862 Lincoln signed an act making Treasury notes legal tender. This was followed by the issue of greenbacks, so called on account of their color, both simple paper and interest-bearing.

The fluctuations in the value of government paper against gold were at times frenzied, depending on the military news, and some serious mistakes were made. In attempts to reduce inflation, Treasury Secretary Chase went in person to the Wall Street markets and sold gold, and he got Congress to pass an act prohibiting contracts in gold on pain of fines and imprisonment. This crude and brazen attempt to interfere with the market proved disastrous. Chase was forced to resign, and his successor, William P. Fessenden (1806–69), quickly persuaded Congress to withdraw it. But on the whole inflation was kept under control and some of the wartime measures—the transformation of 1,400 state banks of issue into a much smaller number of national banks, 1863–64, for instance—were highly beneficial and became permanent.[92]

The problem was generals who would fight—and win. General Scott, head of the army, was not a man of the highest wisdom, as we have seen; he was also seventy-five and ultra-cautious. The overall strategy he impressed on Lincoln was to use the navy to blockade the Confederacy, the number of vessels being increased from 90 to 650, and to divide the South by pushing along the main river routes, the Mississippi, the Tennessee, and the Cumberland. But there was a desire among lesser generals, especially Confederate ones, to have a quick result by a spectacular victory, or by seizure of the enemy's capital, since both Richmond and Washington were comparatively near the center of the conflict. In July 1861 one of Davis' warriors, General P.G.T. Beauregard (1818–93), a flashy New Orleans aristo of French descent, who had actually fired the first shots at Sumter, pushed towards Washington in a fever of anxiety to win the first victory. He was joined by another Confederate army under General Joseph E. Johnston (1807–91), and together they overwhelmed the Unionist forces of General Irvin McDowell (1818–85) at Bull Run, July 21, 1861, though not without considerable difficulty. The new Unionist troops ended by running in panic, but the Confederates were too exhausted to press on to Washington.

The battle had important consequences nonetheless. McDowell was superseded by General George B. McClellan (1826–85), a small, precise, meticulous, and seemingly energetic man who knew all the military answers to everything. Unfortunately for Lincoln and the North, these answers added up to reasons for doing nothing, or doing little, or stopping doing it halfway. His reasons are always the same; not enough men, or supplies, or artillery. As the North's overwhelming preponderance in manpower and hardware began to build up, McClellan refused to take advantage of it, by enticing the South into a major battle and destroying its main army. The war secretary said of him and his subordinates: "We have ten generals there, every one afraid to fight . . . If McClellan had a million men, he would swear the enemy had two million, and then he would sit down in the mud and yell for three."[93] Lincoln agreed: "The general impression is daily gaining ground that [McClellan] does not intend to do anything." At one point Lincoln seems to have seriously believed McClellan was guilty of treason and accused him to his face, but backed down at the vehemence of the general's response. Later, he concluded that McClellan was merely guilty of cowardice. When Lincoln visited the troops with his friend O. M. Hatch, and saw the vast array from a high point, he whispered: "Hatch—Hatch, what is all this?" Hatch: "Why, Mr Lincoln, this is the Army of the Potomac." Lincoln (loudly): "No, Hatch, no. This is *General McClellan's bodyguard*."

The best thing to be said for McClellan is that he had close links with Allan Pinkerton (1819–84), the Scots-born professional detective, who had opened a highly successful agency in Chicago. During Lincoln's campaign for the presidency, and his inauguration, Pinkerton had organized his protection, and undoubtedly frustrated at least one plot to assassinate him. McClellan employed him to build up a system of army intelligence, part of which worked behind Confederate lines, with great success. It eventually became the nucleus of the federal secret service. But Lincoln seems to have known little of this. He believed, almost certainly rightly, that at Antietam in September 1862, McClellan, with his enormous preponderance, could have destroyed the main Confederate army, had he followed up his initial successes vigorously, and thus shortened the war. So he finally removed his non-fighting general, and Pinkerton went with him; and the absence of Pinkerton's thoroughness was the reason why it proved so easy to murder Lincoln in 1865.[94]

First Bull Run had mixed results for the Confederates. It appeared to be the doing of Beauregard, and so thrust him forward: but he proved one of the least effective and most troublesome of the South's generals. In fact the victory was due more to J. E. Johnston, who was a resolute, daring, and ingenious army commander. On April 6–7, 1862, in the first major battle of the war at Shiloh, at Pittsburg Landing in Tennessee, A. S. Johnston hurled his 40,000 troops against General Ulysses S. Grant (1822–85), who had only 33,000. The first day's fighting brought overwhelming success to the Confederates but A. S. Johnston was wounded towards the end of it. That proved a disaster for the South: nor only was their best general to date lost, but Grant turned the tide of battle the next day by leading a charge personally and the Confederates were routed. However, J. E. Johnston was not the only man brought to the fore by First Bull Run. During the melée, the officer commanding the South Carolina volunteers rallied his frightened men by pointing to the neighboring brigade commanded by General Thomas J. Jackson (1824–63) and saying: "There stands Jackson like a stone wall." The name stuck and Jackson's fame was assured. But it was inappropriate. Jackson was not a defensive commander but a most audacious and determined offensive one, with the true killer instinct of a great general. There was only one way the South might win the war. That was by enveloping and destroying in battle the main Unionist Army of the Potomac, taking Washington and persuading the fainthearts on the Unionist side— there were plenty of them—that the cost of waging the war was too high and that a compromise must be sought. Had Lincoln thus been deserted by a majority in Congress, he would have resigned, and the whole of American history would have been different.[95]

Jackson was an orphan, the son of a bankrupt lawyer from Allegheny, Virginia. He was about as unSouthern as it was possible for a Virginia gentleman to be. As Grant put it, "He impressed me always as a man of the Cromwell stamp, much more of a New Englander than a Virginian." He was a Puritan. There is a vivid pen-portrait of him by Mrs James Chesnut, a Richmond lady who kept a war diary. He said to her dourly: "I like strong drink— so I never touch it." He sucked lemons instead and their sourness pervaded his being. He had no sense of humor, and tried to stamp out swearing and obscene joking among his men. He was "an ungraceful horseman mounted on a sorry chestnut with a shambling gait, his huge feet with out-turned toes thrust into his stirrups, and such parts of his countenance as the low visor of his stocking cap failed to conceal wearing a wooden look." Jackson had no slaves and there are grounds for believing he detested slavery. In Lexington he set up a school for black children, something most Southerners hated—in some states it was unlawful—and persisted in it, despite much cursing and opposition. His sister-in-law, who wrote a memoir of him, said he accepted slavery "as it existed in the Southern States, not as a thing desirable in itself, but as allowed by Providence for ends it was not his business to determine."

Yet, as Grant said, "If any man believed in the rebellion, he did." Jackson fought with a ferocity and single-minded determination which no other officer on either side matched. Mrs Chesnut records a fellow-general's view: "He certainly preferred a fight on Sunday to a sermon. [But] failing to manage a fight, he loved next best a long, Presbyterian sermon, Calvinist to the core. He had no sympathy for human infirmity. He was the true type of all great soldiers. He did not value human life where he had an object to accomplish." His men feared him: "He gave orders rapidly and distinctly and rode away without allowing answer or remonstrance. When you failed, you were apt to be put under arrest." He enjoyed war and battle, believing it was God's work, and he was ambitious in a way unusual for Southerners, who were happy-go-lucky except in defense of their beliefs and ways. Jackson would have liked to have been a dictator for righteousness. But, having won the terrifying Battle of Chancellorsville in May 1863, he was shot in the back by men of one of his own brigades, Malone's, who supposedly mistook him in the moonlight for a Yankee. After Jackson's death the Confederacy lost all its battles except Chickamauga.[96]

Jackson was not the only superb commander on the Confederate side. Colonel John Singleton Mosby (1833–1916), who worked behind the Unionist lines, also had the killer instinct. Like many Southern officers, he was a wonderful cavalryman, but he had solid sense too. General Richard Taylor, son of President Taylor, who wrote the best book about the war from inside the Southern high ranks, summed it up: "Living on horseback, fearless and dashing, the men of the South afforded the best possible material for cavalry. They had every quality but discipline."[97] Mosby would have none of that nonsense and was the first cavalryman to throw away his saber as useless and pack two pistols instead. He hated the Richmond set-up— "Although a revolutionary government, none was ever so much under the domination of red tape as Richmond"—and that was one reason he chose the sabotage role, remote from the order-chattering telegraph. The damage he did to the Unionist lines of communication was formidable and he was hated accordingly. On Grant's orders, any of his men who were captured were shot. In the autumn of 1864, for instance, General George Custer executed six of them: he shot three, hanged two, and a seventeen-year-old boy, who had borrowed a horse to join

Mosby, was dragged through the streets by two men on horses and shot before the eyes of his mother, who begged Custer to treat the boy as a prisoner-of-war. This treatment stopped immediately when Mosby began to hang his prisoners in retaliation.

Mosby was "slender, gaunt and active in figure . . . his feet are small and cased in cavalry boots with brass spurs, and the revolvers in his belt are worn with an air of 'business.'"[98] He had piercing eyes, a flashing smile, and laughed often but was always in deadly earnest when fighting. He was the stuff of which Hollywood movies are made and indeed might have figured in one since he lived long enough to see *Birth of a Nation*. He became a myth-figure in the North: he was supposed to have been in the theater when Lincoln was shot, master-minding it, and to have planned all the big railroad robberies, long after the war. But he was the true-life hero of one of the best Civil War stories. During a night-raid he caught General Edwin H. Stoughton naked in bed with a floozie and woke him up roughly. "Do you know who I am, sir?" roared the general. Mosby: "Do you know Mosby, General?" Stoughton: "Yes! Have you got the — — rascal?" Mosby: "No, but *he has got you!*"[99]

Jackson and Mosby were two of the few Confederate leaders who were consistently successful. Jackson's death made it inevitable that Lee would assume the highest command, though it is only fair to Lee to point out that he was finally appointed commander-in-chief of the Southern forces only in February 1865, just two months before he was forced to surrender them at Appomattox. Lee occupies a special place in American history because he was the South's answer to the North's Lincoln: the leader whose personal probity and virtuous inspiration sanctified their cause.[100] Like Lincoln, though in a less eccentric and angular manner, Lee looked the part. He radiated beauty and grace. Though nearly six feet, he had tiny feet and there was something feminine in his sweetness and benignity. His fellow-cadets at West Point called him the "Marble Model." With his fine beard, tinged first with gray, then white, he became a Homeric patriarch in his fifties. He came from the old Virginian aristocracy and married into it. His father was Henry Lee III, Revolutionary War general, Congressman and governor of Virginia. His wife, Anne Carter, was great-granddaughter of "King" Carter, who owned 300,000 acres and 1,000 slaves. That was the theory, anyway. In fact Lee's father was also "Light Horse Harry," a dishonest land-speculator and bankrupt, who defrauded among others George Washington. President Washington dismissed his claim to be head of the United States Army with the brisk, euphemistic, "Lacks economy." Henry was jailed twice and when Robert was six fled to the Caribbean, never to return. Robert's mother was left a penurious widow with many children and the family's reputation was not improved by a ruffianly stepson, "Black Horse Harry," who specialized in adultery.

So Lee set himself quite deliberately to lead an exemplary life and redeem the family honor. That was a word he used often. It meant everything to him. He led a blameless existence at West Point and actually saved from his meager pay at a time when Southern cadets prided themselves on acquiring debts. His high grades meant he joined the elite Corps of Engineers in an army whose chief occupation was building forts. He worked on taming the wild and mighty river Mark Twain described so well. Lee served with distinction in the Mexican War, ran West Point, then commanded the cavalry against the Plains Indians. It was he who put down John Brown's rebellion and reluctantly handed him over to be hanged. He predicted from the start that the "War between the States," as the South called, and calls, it, would be long and bloody. All his instincts were eirenic and, the son of an ardent federalist, he longed for a compromise which would save the Union. But, as he watched the Union Washington had created fall apart, he clung to the one element in it which seemed permanent—Virginia, from which both he and Washington had come and to which he was honor-bound. As he put it, "I prize the Union very highly and know of no personal sacrifice I would not make to preserve it, *save that of honor.*"[101]

Lee was a profound strategist who believed all along that the South's only chance was to entrap the North in a decisive battle and ruin its army. That is what he aimed to do. With Johnston's death he was put in command of the Army of Northern Virginia and ran it for the next three years with, on the whole, great success. He ended McClellan's threat to Richmond (insofar as it was one) in the Seven

Days Battle, routed the Unionists at Second Bull Run (August 1862) but was checked at Antietam the following month. He defeated the Unionists again at Fredericksburg in December 1862 and again at Chancellorsville in May 1863. This opened the way for an invasion of Pennsylvania, heart of the North's productive power, which would force it to a major battle. That is how Gettysburg (July 1863) came about. It was what Lee wanted, an encounter on the grandest possible scale, though the actual meeting-point was accidental, both Lee and General George G. Meade (1815–72), the Unionist commander, blundering into it. Lee had strategic genius, but as field commander he had one great weakness. His orders to subordinate generals were indications and wishes rather than direct commands. As his best biographer has put it, "Lee was a soldier who preferred to suggest rather than order, a general who attempted to lead from consensus and shrank from confrontation. He insisted on making possible for others the freedom of thought and action he sought for himself." This method of commanding a large army sometimes worked for Lee but at Gettysburg it proved fatal. On the first day the Confederate success was overwhelming, and on the second (July 2), General James Longstreet (1821–1904) led the main attack on the Union right but delayed it till 4 P.M. and so allowed Meade to concentrate his main force on the strongpoint of Cemetery Ridge. Some positions were secured, however, including Culp's Hill. Meade's counterattack on the morning of July 3 retook Culp's Hill and confronted Lee with the crisis of the battle. He ordered an attack on Cemetery Ridge but did not make it clear to Longstreet that he wanted it taken at any cost. Jackson would have made no bones about it—take the hill or face court-martial. The charge was led by the division commanded by General George E. Pickett (1825–75), with a supporting division and two further brigades, 15,000 in all. Longstreet provided too little artillery support and the assault force was massacred by enfilading Union artillery, losing 6,000 men. Only half a company of Pickett's charge reached the crest; even so, it would have been enough, and the battle won, if Longstreet had thrown in all his men as reinforcements. But he did not do so and the battle, the culmination of the Civil War on the main central front, was lost. Lee sacrificed a third of his men and the Confederate army was never again capable of winning the war. "It has been a sad day for us," said Lee at one o'clock the next morning, "almost too tired to dismount." "I never saw troops behave more magnificently than Pickett's division . . . And if they had been supported as they were to have been—but for some reason not yet fully explained to me, were not—we would have held the position and the day would have been ours." Then he paused, and said "in a loud voice": "Too bad! *Too bad!* OH! TOO BAD!"[102]

General Meade was criticized for not following up Lee's retreating forces immediately and with energy, but that was easier said than done—his own men had been terribly mauled. But he was a reliable general and with him in charge of the main front on the Atlantic coast Lincoln could be satisfied. Meanwhile, the war in the West was at last going in the Union's favor. Lincoln's strategy was to neutralize as much of the South as he could, divide it and cut it into pieces, then subdue each separately. The naval war, despite the North's huge preponderance in ships, did not always go its way. The South equipped commercial raiders who altogether took or sank 350 Northern merchant ships, but this was no more than minor attrition. When the Union forces abandoned the naval yard at Portsmouth, Virginia, at the beginning of the war, they scuttled a new frigate *Merrimac*. The Confederates raised it, renamed it *Virginia*, and clad it in iron. It met the Union ironclad *Monitor* in Hampton Roads on March 9, 1862 in an inconclusive five-hour duel, the first battle of iron ships in history. But the Confederates were not able to get the *Virginia* into the Mexican Gulf, where it might have served a strategic purpose. They stationed more troops guarding its base than it was worth. The South could run the blockade but they never came near breaking it, and the brilliant campaign of Commodore David Farragut in the Gulf finally sealed the mouth of the Mississippi.

To the north, and in the Western theater, General Grant achieved the first substantial Union successes on land when he took Forts Henry and Donelson; and after Shiloh he commanded the Mississippi as far south as Vicksburg. The North now controlled the Tennessee River and the Cumberland and it took New Orleans and Memphis. But the South still controlled 200 miles of the Mississippi between Vicksburg and Port Hudson, Louisiana. Vicksburg was strongly fortified and protected by natural defenses. Attempts to take it, in May–June 1862 and again in December–January 1863 failed. In May 1863 Grant made a third attempt, and after a fierce siege in which each side lost 10,000, he forced it to surrender the day after Meade won Gettysburg (July 4). Five days later Port Hudson fell, the entire Mississippi was in Union hands, and the Confederacy was split in two.

In Grant Lincoln at last found a war-winning general, and a man he could trust and esteem. Unlike the others, Grant asked for nothing and did not expect the president to approve his plans in advance and so take the blame if things went wrong.[103] Grant was an unprepossessing general. Lincoln said: "He is the quietest little man you ever saw. He makes the least fuss of any man I ever knew. I believe on several occasions he has been [in my office] a minute or so before I knew he was there. The only evidence you have that he's in any particular place is that he makes things move." Grant was born in 1822 at Point Pleasant, Ohio. His father was a tanner. In his day West Point was, as he put it, a place for clever, hardworking boys "from families that were trying to gain advancement in position or to prevent slippage from a precarious place."[104]

Lee, an aristocrat of sorts, was unusual. With Grant at West Point were Longstreet, McClellan, and Sherman, among other Civil War generals—all of them meritocrats. The chief instructor in Grant's day, Denis Hart Mahan—father of the outstanding naval strategist—taught them that "carrying the war into the heart of the assailant's country is the surest way of making him share its burdens and foil his plans." Lee was never able to do this— Grant and Sherman did. Grant was in the heat of the Mexican War, fighting at Palo Alto, Resaca, Monterrey, and Mexico City, and he learned a lot about logistics, later his greatest strength. But he hated and deplored the war, which he regarded as wholly unjust, fought by a Democratic administration in order to acquire more slave states, especially Texas. He saw the Civil War as a punishment on the entire country by God—"Nations, like individuals, are punished for their transgressions. We got our punishment in the most sanguinary and expensive war of modern times."

Grant was a man with a strong and simple moral sense. He had a first-class mind. He might have made a brilliant writer—both his letters and his autobiography have the marks of genius. He made an outstanding soldier. But there were fatal flaws in his system of self-discipline. All his adult life he fought a battle with alcohol, often losing it. After the Mexican War, in civilian life, he failed as a farmer, an engineer, a clerk, and a debt-collector. In 1861 he was thirty-nine, with a wife, four children, a rotten job, and not one cent to his name, in serious danger of becoming the town drunk. He welcomed the Civil War because he saw it as a crusade for justice. It changed his life. A neighbor said: "I saw new energies in him. He dropped his stoop-shouldered way of walking and set his hat forward

on his forehead in a jaunty fashion." He was immediately commissioned a colonel of volunteers and,
shortly after, brigadier-general. He was not impres-
sive to look at. He was a small man on a big horse,
with an ill-kept, scrappy beard, a cigar clamped
between his teeth, a slouch hat, an ordinary soldier's
overcoat. But there was nothing slovenly about his
work. He thought hard. He planned. He gave clear
orders and saw to it they were obeyed, and followed up. His handling of movements and supplies
was always meticulous. His Vicksburg campaign,
though daring, was a model of careful planning,
beautifully executed. But he was also a killer. A nice
man, he gave no mercy in war until the battle was
won. Lincoln loved him, and his letters to Grant are
marvels of sincerity, sense, brevity, fatherly wisdom,
and support. In October 1863 Lincoln gave Grant
supreme command in the West, and in March 1864
put him in charge of the main front, with the title of
General-in-Chief of the Union army and the rank
of lieutenant-general, held by no one since Washington and specially revived in Grant's favor by a
delighted Congress.[105]

Nevertheless, the war was not yet won, and it
is a tribute to the extraordinary determination
of people in the South, and the almost unending
courage of its soldiers, that, despite all the South's
handicaps, and the North's strength, the war continued into and throughout 1864, more desperate than ever. The two main armies, the Army of
the Potomac (North) and the Army of Northern
Virginia (South) had faced each other and fought

each other for three whole years and, as Grant said, "fought more desperate battles than it probably ever before fell to the lot of two armies to fight, without materially changing the vantage ground of either"—it was, indeed, a murderous foretaste of the impenetrable Western Front of World War One. What to do, then? Grant, after much argument with Lincoln, who steered him away from more ambitious alternatives, determined on a two-pronged strategy. One army under General William T. Sherman (1820–91), who had taken over from Grant as commander-in-chief in the West, would sweep through Georgia and destroy the main east–west communications of the Confederacy. Grant's main army would clear the almost impassable Wilderness Region west of Fredericksburg, Virginia, in preparation for a final assault on Lee's army. The Battle of the Wilderness began on May 5–6, 1864, while on the 7th Sherman launched his assault on Atlanta and so to the sea.

The Wilderness battle proved indecisive, though horribly costly in men, and three days later Grant was repulsed at Spotsylvania with equally heavy loss. At the end of the month Grant again attacked at Cold Harbor, perhaps the most futile slaughter of the entire war. In six weeks Grant had lost 60,000 men. Lee, too, had lost heavily—20,000 men, which proportionate to his resources was even more serious than the North's casualties. Nonetheless, Lincoln was profoundly disturbed by the carnage and failure. The Speaker of the House, Schuyler Colfax, found him pacing his office, "his

long arms behind his back, his dark features con-
tracted still more with gloom," explaining: "Why
do we suffer reverses after reverses? Could we
have avoided this terrible, bloody war? . . . Is it ever
to end?" Francis B. Carpenter, who was painting
his *First Reading of the Emancipation Proclamation by
President Lincoln,* described him in the hall of the
White House, "clad in a long morning wrapper,
pacing back and forth a narrow passage leading to
one of the windows, his hands behind him, great
black rings under his eyes, his head bent forward
upon his breast—altogether . . . a picture of the
effects of sorrow, care and anxiety."[106]

All the same, the noose was tightening round
the South. Davis himself felt it. Even before Gettysburg, he had personally been forced to quell a
food riot of hungry women in Richmond. Unionist troops overran his and his brother's property,
taking the whites prisoner and allowing the blacks
to go. Some 137 slaves fled to freedom leaving, on
Davis' own estate, only six adults and a few children. His property was betrayed by a slave he
trusted, the soldiers cut his carpets into bits as sou-
venirs, they drank his wine, stabbed his portrait
with knives, and got all his private papers, spicy
extracts from which duly appeared in the Northern newspapers. In Richmond, Davis had to sell
his slaves, his horses, and his carriage just to buy
food—ersatz coffee, pones or corncakes, bread, a
bit of bacon. Jeb Stuart, Davis' best cavalry commander, fell, mortally wounded. He had one
good general, Lee, marking Grant; but Lincoln

had two—and Sherman now took Atlanta, moved through Georgia, burning and slaughtering, and on December 21, 1864 was in Savannah, having cut the Confederacy in two yet again. By Christmas much of the South was starving. Davis had made Lincoln's job of holding the North together easier by proclaiming, for four years, that he would not negotiate about anything except on the basis of the North admitting the complete independence of the South. Now he again insisted the South would "bring the North to its knees before next summer." On hearing this rodomontade, his own vice-president, Alexander Stephens (1812–83), told him in disgust he was leaving for his home and would not return—it was the beginning of the disintegration of the Confederate government.[107]

Much of the South was now totally demoralized by military occupation. Sarah Morgan of Baton Rouge, who kept a diary, described the sacking of her house:

one scene of ruin. Libraries emptied, china smashed, sideboards split open with axes, three cedar chests cut open, plundered and set up on end; all parlor ornaments carried off. [Her sister Margaret's] piano, dragged to the center of the parlor had been abandoned as too heavy to carry off; her desk lay open with all letters and notes well thumbed and scattered around, while Will's last letter to her was open on the floor, with the Yankee stamp of dirty fingers. Mother's portrait half cut from the frame stood on the floor. Margaret, who was present at the sacking, told how she had saved father's. It seems that those who wrought destruction in our house were all officers![108]

The destruction in Georgia was worse. Like Grant, Sherman was a decent man but a fierce, killer general, determined to end the war and the slaughter as speedily as possible and, with this his end, anxious to demonstrate to the South in as plain a manner as he could that the North was master and resistance futile. He cut a swathe 60 miles wide through Georgia, destroying everything— railroads, bridges, crops, cattle, cotton-gins, mills, stocks—which might conceivably be useful to the South's war-effort. Despite his orders, and the generally tight discipline of his army in action, the looting was appalling and the atrocities struck fear and dismay into the stoutest Southern hearts.[109]

Sherman's capture of Atlanta and his rout of the Southern army in Georgia came in time—just— to insure Lincoln's reelection. During the terrible midsummer of 1864 there had been talk, by "Peace Democrats," of doing a deal with Davis and getting control of both armies, thus ending both the rebellion and Republican rule. Many prominent Republicans thought the war was lost and wanted to impose Grant as a kind of president–dictator. He wrote to a friend saying he wanted "to stick to the job I have"—and the friend showed it to Lincoln. Lincoln observed: "My son, you will never know how gratifying that is to me. No man knows, when that presidential grub starts to gnaw at him, just how deep it will get until he has tried it. And I didn't know but what there was one gnawing at Grant." The general put an end to intrigue by stating: "I consider it as important to the cause that [Lincoln] should be reelected as that the army should be successful in the field."[110]

Sherman's successes in September, and his continued progress through Georgia, swung opinion strongly back in Lincoln's favor. The increasing desperation of the South, expressed in terrorism, bank-raids, and murder in Northern cities, inflamed the Northern masses and were strong vote-winners for the Republicans. The resentful McClellan fared disastrously for the Democrats. Lincoln carried all but three of the participating states and 212 electoral votes out of 233, a resounding vote of confidence by the people.[111] He entered his second term of office in a forthright but still somber mood, in which the religious overtones in his voice had grown stronger. They echo through his short Second Inaugural, a meditation on the mysterious way in which both sides in the struggle invoked their God, and God withheld his ultimate decision in favor of either:

Both read the same Bible, and pray to the same God; and each invokes His aid against the other. It may seem strange that any men should dare to ask a just God's assistance in wringing their bread from the sweat of other men's faces; but let us judge not that we be not judged. The prayers of both could not be answered; that of neither has been answered fully. The Almighty has His own purposes: "Woe unto the world because of offenses! for it needs be that offenses come, but woe unto that man by whom the offenses cometh!" . . . Fondly do we hope—fervently do we pray— that this mighty scourge of war may speedily pass away. Yet if God wills that it continue, until all the wealth piled by the bond-man's two-hundred-and-fifty years of unrequited toil shall be sunk, and until every drop of blood drawn by the lash shall be paid with another drawn by the sword, as was said three thousand years ago, so still it must be said "the judgements of the Lord are true and righteous altogether."

So Lincoln asked the nation to continue the struggle to the end, "With malice to none, with charity to all, with firmness in the right, as God gives us to see the right."[112]

The Second Inaugural began the myth of Lincoln in the hearts of Americans. Those who actually glimpsed him were fascinated by his extraordinary appearance, so unlike the ideal American in its massive lack of beauty, so incarnate of the nation's spirit in some mysterious way. Nathaniel Hawthorne wrote (1862):

The whole physiognomy is as coarse a one as you would meet anywhere in the length and breadth of the state; but withal, it is redeemed, illuminated, softened and brightened by a kindly though serious look out of his eyes, and an expression of homely sagacity, that seemed weighted with rich results of village experience. A great deal of native sense, no bookish cultivation, no refinement; honest at heart, and thoroughly so, and yet in some sort, sly—at least endowed with a sort of tact and wisdom that are akin to craft, and would impel him, I think, to take an antagonist in flank, rather than make a bull-run at him right in front. But on the whole I like this sallow, queer, sagacious visage, with the homely human sympathies that warmed it; and, for my small share in the matter, would as lief have Uncle Abe for a ruler as any man that it would have been practical to have put in his place.[113]

Walt Whitman, looking at the president from a height in Broadway, noted "his perfect composure and coolness—his unusual and uncouth height, his dress of complete black, stovepipe hat pushed back on the head, dark-brown complexion, seam'd and wrinkled yet canny-looking face, black, bushy head of hair, disproportionately long neck, and his hands held behind him as he stood observing the people." Whitman thought "four sorts of genius" would be needed for "the complete lining of the Man's future portrait"—"the eyes and brains and finger-touch of Plutarch and Aeschylus and Michelangelo, assisted by Rabelais."[114]

There is a famous photograph of Lincoln, taken at this time, visiting the HQ of the Army of the Potomac, standing with some of his generals outside their tents. These officers were mostly tall for their times but Lincoln towers above them to a striking degree. It was as if he were of a different kind of humanity: not a master-race, but a higher race. There were many great men in Lincoln's day—Tolstoy, Gladstone, Bismarck, Newman, Dickens, for example—and indeed master spirits in his own American—Lee, Sherman, Grant, to name only three of the fighting men—yet Lincoln seems to have been of a different order of moral stature, and of intellectual heroism. He was a strong man, and like most men quietly confident of their strength, without vanity or self-consciousness—and also tender. Towards the end of the war, Lincoln went to see Seward, his secretary of state, a man with whom he often disagreed and whom he did not particularly like. Seward had somehow contrived to break both his arm and his jaw in a carriage accident. Lincoln found him not only bedridden but quite unable to move his head. Without a moment's hesitation, the president stretched out at full length on the bed and, resting on his elbow, brought his face near Seward's, and they held an urgent, whispered consultation on the next steps the administration should take. Then Lincoln talked quietly to the agonized man until he drifted off to sleep. Lincoln could easily have used the excuse of Seward's incapacity to avoid consulting him at all. But that was not his way. He invariably did the right thing, however easily it might be avoided. Of how many other great men can that be said?[115]

Lincoln was well aware of the sufferings of those in the North who actively participated in the struggle. They haunted him. He read to his entourage that terrible passage from *Macbeth* in which the

King tells of his torments of mind:

> *we will eat our meal in fear, and sleep*
> *In the affliction of these terrible dreams,*
> *That shake us nightly; better be with the dead*
> *Than on the torture of the mind to lie*
> *In restless ecstasy.*[116]

One man who was also well aware of the suffering was Whitman. Too old to fight, he watched his younger brother George, a cabinetmaker, enlist for a 100-day stint which turned into four years, during which time he participated in twenty-one major engagements, saw most of his comrades killed, and spent five months in a horrific Confederate prison. Some 26,000 Union soldiers died in these dreadful stockades, and so great was the Union anger at conditions in them, especially at Andersonville, that its commandant, Major Henry Wirz, was the only Southerner to be punished by hanging.[117] Instead of enlisting, Whitman engaged himself in hospital service, first at the New York Hospital, then off Broadway in Pearl Street, later in Washington, D.C.: "I resigned myself / To sit by the wounded and soothe them, or silently watch the dead."

In some ways the Civil War hospitals were bloodier than the battlefield. Amputation was "the trade-mark of Civil War surgery." Three out of four operations were amputations. At Gettysburg, for an entire week, from dawn till twilight, some surgeons did nothing but cut off arms and legs. Many of these dismemberments were quite unnecessary, and the soldiers knew it. Whitman was horrified by what happened to the wounded, often mere boys. He noted that the great majority were between seventeen and twenty. Some had pistols under their pillows to protect their limbs. Whitman himself was able to save a number by remonstrating with the surgeons. He wrote:

> From the stump of the arm, the amputated hand,
> I undo the clotted lint, remove the slough, wash off the
>   matter and blood.
> Back on his pillow the soldier bends with curv'd neck and
>   side-falling head,
> His eyes are closed, his face is pale, he dares not look on
>   the bloody stump,
> And has not yet looked on it.

More arms and legs were chopped off in the Civil War than in any other conflict in which America has ever been engaged—but a few dozen fewer than might have been, but for Whitman, A paragraph in the *New York Tribune* in 1880 quoted a veteran pointing to his leg: "This is the leg [Whitman] saved for me."[118]

Whitman calculated that, during the war, he made over 600 hospital visits or tours, some lasting several days, and ministered in one way or another to over 100,000 soldiers. His book of poems *Drum-Taps* records some of his experiences. Not everyone welcomed his visits. One nurse at the Armory Square hospital said: "Here comes that odious Walt Whitman to talk evil and unbelief to my boys." The scale of the medical disaster almost overwhelmed him—one temporary hospital housed 70,000 casualties at one time. Whitman considered the volume and intensity of the suffering totally disproportionate to any objective gained by the war. Others agreed with him. Louisa May Alcott (1832–88), later author of the famous bestseller *Little Women* (1868), spent a month nursing in the Washington front-line hospitals before being invalided home with typhoid, and recorded her experiences in *Hospital Sketches* (1863). This is a terrifying record of bad medical practice, of the kind Florence Nightingale had utterly condemned a decade before, including lethal overdosing with the emetic calomel. At many points her verdict and Whitman's concurred.[119]

Yet it is curious how little impact the Civil War made upon millions of people in the North. When Edmund Wilson came to write his book on the conflict, *Patriotic Gore: Studies in the Literature of the American Civil War* (1962), he was astonished by how little there was of it. There were hymn-songs, of course: "John Brown's Body," Julia Ward Howe's "Battle Hymn of the Republic," to rally Northern spirits, Daniel Decatur Emmett's "Dixie" to enthuse the South. The young Henry James was not there—he had "a mysterious wound," which prevented serving. Mark Twain was out west. William Dean Howells was a consul in Italy. It was quite possible to live in the North and have no contact with the struggle whatsoever. It is a notable fact that Emily Dickinson (1830–86), America's greatest poet, lived quietly throughout the war in Amherst without it ever impinging on her consciousness, insofar as that is reflected in her poetry. Of her more than 1,700 poems, not one refers directly to the war, or even indirectly, though they often exude terror and dismay. She was educated at Amherst Academy and spent a year at Mount Holyoke Female Seminary: otherwise her life was passed at home, eventless, and for the last twenty-five years of her life in almost complete seclusion. Only six of her poems were published in her lifetime and evidently she did not consider it part of the poet's job to obtain publication. Effectively, she did not emerge as a writer at all until the 1890s, after her death. In a sense, her poetry is internal exploration and could have been written in almost any country, at almost any period of history, with one exception: the South in the 1860s. Had she lived in, say, Charleston or Savannah, she would have been forced to confront external reality in her verse. That is the difference between North and South.[120]

But not only in cloistered New England was the war distant. In vast stretches of America, it had virtually no effect on the rapid development of the country. Not that Westerners were indifferent to the war. They favored the Union because they needed it. The South was protesting not only against the North's interference in its "peculiar institution" but against the growth of government generally. But Westerners, for the time being at least, wanted some of the services that only federal government could provide. As the historian of the trails through Oregon and to California put it, "Most pre-Civil War overlanders found the United States government, through its armed forces, military installations, Indian agents, explorers, surveyors, road builders, physicians and mail-carriers to be an impressively potent and helpful force."[121] Up to the outset of the Civil War, 90 percent of the U.S. Army's active units were stationed in the seventy-nine posts of the trans-Mississippi—7,090 officers and men in 1860. Withdrawal of many units once the war began made Westerners realize quite how dependent they were on federal power.

Lack of troops raised problems in the West and may have encouraged the Indians to take advantage. There were raids and massacres, and the settlers responded by raising volunteers and using them. They were less experienced at dealing with Indians than the regular units, and their officers were often prone to take alarm needlessly and overreact—all the good commanders were out east, fighting. What was liable to happen was demonstrated at Sand Creek in the Colorado Territory on November 29, 1864, just after Lincoln's reelection. Following Indian atrocities, a punitive column consisting of the Third Colorado Volunteers, under Colonel John M. Chivington, attacked a camp of 500 Cheyennes. Their leaders, Black Kettle and White Antelope, believed a peace treaty was in effect and said they had turned in their arms. The volunteers slaughtered men, women, and children indiscriminately, killing over 150 and returning to Denver in triumph, displaying scalps and severed genitals like trophies. This Sand Creek massacre was later investigated by a joint Committee of Congress, and Chivington condemned, though he was never punished. The Cheyennes retaliated brutally on several occasions, and on December 21, 1866, after the war was over, in combination with Lakotas and Arapahos, they ambushed and slaughtered eighty men under the command of Colonel William J. Fetterman, one of the worst defeats the U.S. Army suffered at Indian hands.[122]

In some ways the Civil War hastened the development of the West because, by removing the Southern–Democratic majority in both Houses of Congress, it ended a legislative logjam which had held up certain measures for decades and impeded economic and constitutional progress. For instance, the Californian engineer–promoter Theodore D. Judah, representing a group of San Francisco bankers and entrepreneurs, contrived in the spring of 1861, immediately after the Southerners had left Washington, to lobby the Pacific Railroad Act through Congress. This was entirely a venture to benefit the North and the Northwest. It involved the railroads receiving from the federal government a 400-foot right of way, ten alternate sections of land for each mile of track, and first-mortgage loans of $16,000 per mile in flat country, $32,000 in foothills, and $48,000 per mile in the mountains—an enormous federal subsidy, in effect, which would only have passed over Southern dead bodies. In the event, the subsidy did not prove enough for this giant undertaking and was increased by a further Act of Congress in 1864, the Southerners still being absent.

In fact the North and West got their revenge, during the Civil War, for the many defeats they had suffered at the hands of Southern legislators in the thirty-two years 1829–60. By 1850 the Southern plantation interests had come to see the cheapland policy in the West as a threat to slavery. Their senators killed the Homestead Bill of 1852, and in 1860, after Southerners made unavailing efforts to kill a similar bill, President Buchanan vetoed it. Thus a Homestead Bill became an important part of Lincoln's platform and in 1862 it marched triumphantly through Congress. This offered an enterprising farmer 160 acres of public land, already surveyed, for a nominal sum. He got complete ownership at the end of six months on paying $1.50 an acre, or for nothing after five years' residence.[123] This eventually proved one of the most important laws in American history, the consequences of which we will examine shortly. The removal of Southern resistance also speeded up the constitutional development of the West. Kansas entered the Union as a free state in 1861, Nevada in 1864, and Nebraska soon after the end of the war in 1867. Meanwhile the administration extended the territorial system over the remaining inchoate regions beyond the Mississippi. The Dakotas, Colorado, and Nevada territories were organized in 1861, Arizona and Idaho in 1863, and by 1870 Wyoming and Montana had also become formal territories on the way to statehood.[124]

Out west, then, they just got on with it, and made money. The mining boom, which had cost the South any chance of California becoming a slave state, continued and intensified, thus pouring specie into Washington's war-coffers. The classic boomtown, Virginia City, emerged 7,000 feet up the mountains of Nevada, and was immortalized by Mark Twain. The gold and silver were embedded in quartz, and elaborate crushing machinery—and huge amounts of capital—were needed for the big-pay mines, the Ophir, Central, Mexican, Gould, and Curry. Experienced men from Cornwall, Wales, and the German mountains poured in. The Comstock Lode became the great mineralogical phenomenon of the age. It went straight through Virginia City, from north to south, and laboring men earned the amazing wages of $6 a day, working in three shifts, round the clock. So, as Twain wrote, even if you did not own a "piece" of a mine—and few did not—everyone was happy; "Joy sat on every countenance, and there was a glad, almost fierce intensity in every eye, that told of the money-getting schemes that were seething in every brain and the high hope that held sway in every heart. Money was as plentiful as dust; every individual considered himself wealthy and a melancholy countenance was nowhere to be seen."[125]

Any shots fired in these parts had nothing to do with the Civil War but reflected the normal human appetites of greed, lust, anger, and envy.

And, as Mark Twain put it, "the thin atmosphere seemed to carry healing to gunshot wounds, and therefore to simply shoot your adversary through both lungs was a thing not likely to afford you any permanent satisfaction, for he would be nearly certain to be around looking for you within the month, and not with an opera glass, either." The miners, most of whom were heavily armed, chased away any Indians who stood between them and possible bullion, ignoring treaties. Gold was found in 1860 on the Nez Percé Indians' reservation at the junction of the Snake and Clearwater rivers. The superintendent of Indian affairs reported: "To attempt to restrain these miners would be like attempting to restrain the whirlwind." With Washington's attention on the war, protection of reservations had a low priority and the miners did what they pleased. They created the towns of Lewiston, Boise, and in 1864 Helena, Idaho was a mining-created state; so was Montana, formed out of its eastern part, and Wyoming Territory. Nor was gold and silver the only lure—it was at Butte, Montana, that one of the world's great copper strikes was made. The miners were almost entirely young men between sixteen and thirty; the women nearly all whores. But it was creative: seven states, California, Nevada, Arizona, New Mexico, Colorado, Idaho, and Montana owe their origins to mining—and the key formation period, in most cases, was during the Civil War.[126]

It was totally different in the South: there, nothing mattered, nothing could occur, but the war. Concern for the war, anxiety to win the war, was so intense that people forgot what it was really about. Davis himself forgot to the point where he was among the earliest to urge that slaves should be manumitted in return for fighting for the South. Resistance to this idea was, at first, overwhelming, on the ground that blacks would not or could not fight—this despite the fact that 180,000 blacks from the North were enlisted in the Union army and many of them fought very well indeed. Arguing with a senator who was against enlisting blacks at any price, Davis in exasperation declared: "If the Confederacy falls, there should be written on its tombstone, 'Died of a theory.'"[127] As the Union army sliced off chunks of the South and liberated its slaves, many flocked to join the army—apart from anything else, it was the only way they could earn a living. Slavery itself was breaking down, even in those parts of the South not yet under Union rule. Slaves were walking off the plantations more or less as they chose; there was no one to prevent them, and no one to hunt them once they were at liberty. There was no work and no food for them either. So they were tempted to cross the lines and enlist in the Union forces. Hence Davis redoubled his efforts to persuade the Confederate Congress to permit their enlistment. As he put it, "We are reduced to choosing whether the negroes shall fight for us, or against us."

Eventually on March 13, 1865, Congress accepted his arguments, but even then it left emancipation to follow enlistment only with the consent of the owner. Davis, in promulgating the new law, added a proviso of his own making it compulsory for the owner of a slave taken into war service to provide manumission papers. But by then it was all too late anyway. Granted the fact that slaves formed more than a third of the South's population at the beginning of the Civil War, their prompt conscription would have enormously added to the strength of the Confederate armies. And many of them would have been willing to fight for the South, too—after all, it was their way of life as well as that of the whites which was at stake. It is a curious paradox, but one typical of the ironies of history, that black participation might conceivably have turned the scales in the South's favor. But obstinacy and "theory" won the day and few blacks actually got the chance to fight for their homeland.[128]

The end of the Confederacy was pitiful. On April 1, 1865, Davis sent his wife Varina away from Richmond, giving her a small Colt and fifty rounds of ammunition. The next day he had to get out of Richmond himself. He went to Danville, to plan guerrilla warfare. By this point General Lee was already in communication with General Grant about a possible armistice, and had indeed privately used the word "surrender," but he continued to fight fiercely with his army, using it with his customary skills. He dismissed pressure from junior officers to negotiate, and as late as April 8 he took severe disciplinary action against three general officers who, in his opinion, were not fighting in earnest or had deserted their posts. But by the next morning Lee's army was virtually surrounded. He dressed in his best uniform, wearing, unusually for him, a red silk sash and sword. Having heard the latest news of the position of his troops, and the Union forces, he said: "Then there is nothing left me but to go and see General Grant and I would rather die a thousand deaths."[129]

The two generals met at Appomattox Court House, Grant dressed in "rough garb," spattered with mud. Both men were, in fact, carefully dressed for the occasion, as they wished to appear for posterity. The terms were easily agreed, Grant allowing that Southern officers could keep their sidearms and horses. Lee pointed out that, in the South, the enlisted men in the cavalry and artillery also owned their horses. Grant allowed those to be kept too. After Lee's surrender on April 9, Davis hurried to Greensboro to rendezvous with General Johnston's army. But in the meantime Johnston had reached an agreement with General Sherman which, in effect, dissolved the Confederacy. Davis gave the terms to his Cabinet, saying he wanted to reject them, but the Cabinet accepted them. Washington, however, did not, and the South had to be content, in the end, with a simple laying down of arms.[130]

By this time Lincoln was dead. He had summoned Grant to hear his account of the surrender at Appomattox, and he beamed with pleasure when the general told him that the terms had extended not just to the officers but to the men: "I told them to go back to their homes and their families and that they would not be molested, if they did nothing more." Lincoln expected Sherman to report a similar surrender and he told Grant he expected good news as he had just had one of his dreams which portended such. Grant said he described how "he seemed to be in some singular, indescribable vessel and . . . he was moving with great rapidity to an indefinite shore."[131] Lincoln told his wife (April 14), who said to him, "Dear husband, you almost startle me by your great cheerfulness," "And well may I feel so, Mary. I consider *this day,* the war has come to a close."

On April 14 they went to a performance of the comedy *Our American Cousin* at Ford's Theater. Lincoln was no longer protected by Pinkerton, but Marshal Ward Hill Lamon, who often served as his bodyguard, begged him not to go to the theater or any similar place, and on no account to mingle with promiscuous crowds. That evening was particularly dangerous since it had been widely advertised that Grant, too, would join the president in his theater excursion. Name, date, time, place—all were published. John Wilkes Booth (1838–65), from an acting family of British origins, also noted for mental instability, and brother of the famous tragedian Edwin Booth, was a self-appointed Southern patriot. He had three days to organize the assassination, with various associates. He also planned to kill Seward and Vice-President Andrew Johnson (1808–75), regarded with peculiar abhorrence in the South because he was a Democrat and a Southerner, from Tennessee, the only Southerner who remained in the Senate in 1861—and accordingly rewarded with the vice-presidency in Lincoln's second term.

Booth had no difficulty in getting into the theater, and he obtained entry to the president's box simply by showing Charles Forbes, the White House footman on duty, his calling-card. He barred the door of the box, moved behind Lincoln, who was leaning forward, then aimed his Derringer at the back of the president's head and pulled the trigger. He then drew a knife, stabbing Lincoln's ADC, jumped from the box, breaking his ankle in the process, shouted "Sic semper tyrannis," the motto of the State of Virginia, and escaped through the back of the theater. Two weeks later he was shot and killed in Bowling Green, Virginia. Lincoln himself was taken to a nearby house where he lingered for nine hours, never regaining consciousness.

It is clear that Booth had links going back to Richmond but equally clear that Davis knew nothing about the assassination plot and would never have authorized it. But many at the time believed he was involved. His last days of liberty were clouded by rumors, including one that a price of $100,000 was on his head and another that he was dressed as a woman. He was taken on May 10. Almost his last words to his colleagues were that he was glad "no member of his Cabinet had made money out of the war and that they were all broke and poor." He himself gave his last gold coin to a little boy presented to him as his namesake. All he then had in his pockets was a wad of worthless Confederate scrip. His soldiers–captors jeered at him: "We'll hang Jeff Davis from a sour apple tree." Their commander, Major-General James Wilson, said later: "The thought struck me once or twice that he was a mad man."

Davis was put in heavy leg-irons and taken to Fort Monroe, opposite Norfolk, Virginia, where he was held for 720 days mostly in solitary confinement, and subjected to many humiliations, with bugs in his mattress and only a horse-bucket to drink from. None of this would have happened had Lincoln lived. Johnson, now president, insisted on this to prove to Northern opinion that he was not favoring a fellow-Southerner. On the other hand, he hated the idea, put forward by Stanton, the secretary of war, and others, that Davis should be tried, convicted, and hanged. So he allowed Dr. John J. Craven, who visited Davis many times in his cell and had long conversations with him, to smuggle out his diaries and have them written up by a popular writer, Charles G. Halpine. They appeared as *The Prison Life of Jefferson Davis,* presenting him as a tragic hero, aroused much sympathy, even in the North, and prepared the way for his release. Davis detested the book. He refused to ask for a pardon, demanding instead a trial which (he was sure) would lead to his acquittal and vindicate him totally. Instead, a writ of habeas corpus (which Lincoln had suspended but was now permitted again) got him out in May 1867. He then went to Canada, and wrote rambling memoirs, lived to bury all his sons, and died, full of years and honor—in the South at least—in 1889. His funeral, attended by a quarter of a million people, was the largest ever held in the South.[132]

Lee, by contrast, was broken and tired and did not last long. When he died in 1870 people were amazed to learn he was only sixty-three. He spent his last years in the thankless job of running a poor university, Washington College [now Washington and Lee], believing that "what the South needs most is education." He refused to write his memoirs, blamed no one, avoided publicity, and, when in doubt, kept his mouth shut. Legend has it that his last words were "Tell Hill he *must* come up!" and "Strike the tent!" In fact he said nothing.[133]

The end of the Civil War solved the problem of slavery and started the problem of the blanks, which is with America still. Everyone, from Jefferson and Washington onwards, and including Lincoln himself, had argued that the real problem of slavery was not ending it but what to do with the freed blacks afterwards. All these men, and the overwhelming majority of ordinary American whites, felt that it was almost impossible for whites and blacks to live easily together. Lincoln did not regard blacks as equals. Or rather, they might be morally equal but in other respects they were fundamentally different and unacceptable as fellow-citizens without qualification. He said bluntly that it was impossible just to free the slaves and make them "politically and socially our equals." He freely admitted an atti-

tude to blacks which would now be classified as rac-
ism: "My own feelings will not admit [of equality]."
The same was true, he added, of the majority of
whites, North as well as South. "Whether this feeling accords with justice and sound judgment is not
the sole question. A universal feeling, whether well-
or ill-founded, cannot be safely disregarded."[134] He
told a delegation of blacks who came to see him at
the White House and asked his opinion about emigration to Africa or elsewhere, that he welcomed
the idea: "There is an unwillingness on the part of
our people, harsh as it may be, for you free colored
people to remain with us." He even founded an
experimental colony on the shores of San Domingo,
but the dishonesty of the agents involved forced the
authorities to ship the blacks back to Washington.
All schemes to get the blacks back to Africa had
been qualified or total failures, for the simple reason that only a tiny proportion of them ever had the
smallest desire to return to a continent for which,
instinctively, they felt an ancestral aversion. Like
everyone else, they wanted to remain in the United
States, even if life there had its drawbacks.

That being so, what to do? And what to do with
the rebellious South? On November 19, 1863, Lincoln had made a short speech at the dedication of
the cemetery at Gettysburg. It consisted of only
261 words, and it did not make much impact at the
time—the professional orator Edward Everett, for-
merly president of Harvard, was the chief speaker
on the occasion—but its phrases have reverberated
ever since, and the ideas those few short words pro-

jected have penetrated deep into the consciousness of humanity.[135] Lincoln reminded Americans that their country was "dedicated to the proposition that all men were created equal" and that the war was being fought to determine whether a nation so dedicated "can long endure." Second, he referred to "unfinished work" and "the great task remaining before us." This was to promote "a new birth of freedom" in America, by which he meant "government of the people, by the people, for the people." Lincoln, then, thought the blacks should be treated as equals, politically and before the law; but at the same time he insisted that America was a democracy—and Southern whites, rebels though they might be, had as much right to participate in that democracy as the loyalists. How to reconcile the two?

Lincoln's intentions are known because, while still living, he had to deal with the problem of governing those parts of the South occupied by Union armies. He was clear about two things. First, political justice had to be done to the blacks. Second, the South must be got back to normal government as quickly as possible once the spirit of rebellion was exorcized. He proposed a general amnesty, to qualify for which "politically accused persons" would have merely to take an oath to abide by the Constitution. A state government would be valid, and recognized by Washington, if not less than 10 percent of the voters who were on the rolls in 1860, and had taken the loyalty oath, voted for it. He wanted the occupying armies withdrawn as soon as possible, but he wanted the blacks on the voting rolls first: "We must make voters of them before we take away the troops. The ballot will be their only protection after the bayonet is gone." All this was set down in his Proclamation of Amnesty and Reconstruction, issued December 8, 1863.[136]

His first practical step was to get Congress to pass the Thirteenth Amendment. Its first section banned slavery and "involuntary service" (except for crimes, after conviction by due process) anywhere in the United States, "or any place subject to their jurisdiction." Section Two empowered Congress "to enforce this article by appropriate legislation." Lincoln did not live to see the Amendment adopted by the three-quarters majority of the states that it required, but it was clear he was fully committed to the liberation of slaves and to entrusting them with the vote. It was also clear that he was in favor of the spirit of the Fourteenth Amendment, adopted in 1868, which wound up the unfinished business of the Civil War, by dealing with the eligibility for office of former rebels and the debts incurred by the Confederacy, but, above all, by making all born or naturalized citizens of the United States equal politically and judicially, and by making it unconstitutional for any state to "deny to any person within its jurisdiction the equal protection of the laws." This very important constitutional provision carried forward Lincoln's policy of justice to the blacks into the future, and became in time the basis for desegregation in the South.[137]

Balancing this, it was abundantly clear that

Lincoln wanted to exercise the utmost clemency. He intended to bind wounds. On April 14, 1865, his friend Gideon Welles described him as cheerful, happy, hoping for peace, "full of humanity and gentleness." His last recorded words on the subject of what to do with the South and the leaders of the rebellion were: "No one must expect me to take any part in hanging or killing these men, even the worst of them. Frighten them out of the country, open the gates, let down the bars, scare them off. Enough lives have been sacrificed; we must extinguish our resentments if we expect harmony and union. There is too much disposition, in certain quarters, to hector and dictate to the people of the South, to refuse to recognise them as fellow-citizens. Such persons have too little respect for Southerners' rights. I do not share feelings of that kind."[138]

However, Lincoln was dead, and the task of reconstruction fell on his successor, Andrew Johnson. Johnson agreed wholly with Lincoln's view that the South, consistent with the rights of the freed slaves, should be treated with leniency. But he was in a much less strong position to enforce such views. He had not been twice elected on a Northern Republican platform, fought and won a Civil War against the rebels, and held the nation together during five terrifying years. Moreover, he was a Southerner—and, until 1861, a lifelong Democrat. The fact that he had defied the whole might of the Southern establishment in 1861 by being the only Southern senator to remain in Washington when the South seceded was too easily brushed aside. So, too, was his profound belief in democracy. John-son stood for the underdog. He had nothing in common with the old planter aristocracy who had willed the war and led the South to destruction. In many respects he was a forerunner of the Southern populists who were soon to make their entry on to the American scene.

He was born in Raleigh, North Carolina. His background was modest, not to say poor. He seems to have been entirely self-educated. At thirteen he was apprenticed to a tailor but ran away from his cruel master and came to Greeneville, Tennessee, where he plied his trade and eventually became its mayor. He was a typical Jacksonian Democrat, strongly in favor of cheap land for the poor—his passionate belief in the Homestead Act was a major factor in his breach with the Southern leadership in 1860–61. He was state representative and senator and governor, representative and senator in Congress, and finally (in Lincoln's first term) military governor of Tennessee from 1862. He was a brilliant speaker, but crude in some ways, with a vile temper. And he drank. At Lincoln's Second Inaugural, following his own swearing-in, Johnson, who had been consuming whiskey, insisted on making a long, rambling speech, boasting of his plebeian origins and reminding the assembled dignitaries from the Supreme Court and the diplomatic corps, "with all your fine feathers and gewgaws," that they were but "creatures of the people." Lincoln was disgusted and told the parade marshal, "Do not let Johnson speak outside."[139]

Johnson began his term with a violent denunciation of all rebels as "traitors" who "ought to be hanged." Then he proceeded to change tack and carry out what he believed were Lincoln's wishes and policies. There were three possible constitutional positions to be taken up about the South. The extreme position, urged on the White House and Congress by Senator Charles Sumner, the firebrand who had been caned in the Senate, and by Thaddeus Stevens (1792–1868), chairman of the House Ways and Means Committee, was that secession had, in effect, destroyed the Southern states, which now had no constitutional existence, and it was entirely in the power of Congress to decide when and how they were to be reconstituted. Both men were, first and foremost, good haters, and they hated the South and wanted to punish it to the maximum of their power. And their power, in both Houses of Congress, was enormous. Second, there was the bulk of the Republican majority who took a somewhat more moderate position: the rebellion had not destroyed the Southern states but it had caused them to forfeit their constitutional rights, and it was up to Congress to determine when those rights should be restored, under the article of the Constitution guaranteeing all states a republican form of government. Finally there was the Lincoln–Johnson clemency position: this held that rebellion had not affected the states at all, beyond incapacitating those taking part in it from performing their constitutional duties, and that this disbarment could be removed by executive pardon—as soon as this was done, normal government of the states, by the states, could follow.[140]

Initially, Johnson was in a strong position to make this third position prevail. Not only was it manifestly Lincoln's wish, but he was called on to act alone, since it was against the practice of the United States political system for a Congress elected in the autumn of 1864 to be summoned before December 1865, unless by special presidential summons. He had, then, a free hand, but whether it was wise to exercise it without the closest possible consultation with Congressional leaders is doubtful. On May 29, 1865 Johnson issued a new proclamation, extending Lincoln's clemency by excluding from the loyalty oath-taking anyone in the South with property worth less than $20,000. This was consistent with his general view that the South had been misled by its plantocracy and that it must be rebuilt by the ordinary people. In the early summer, he appointed provisional governors for each rebel state, with instructions to restore normalcy as soon as practicable, provided each state government abolished slavery by its own law, repudiated the Confederation's debts, and ratified the Thirteenth Amendment. This was quickly done. Every state found enough conservatives, Whigs, or Unionists, to carry through the program. Every state amended its constitution to abolish slavery. Most repudiated the Confederate debt. All but Mississippi and Texas ratified the Thirteenth Amendment. When all, including these two sluggards, had elected state officials, Johnson felt able to declare the rebellion legally over, in a proclamation dated April 6, 1866.[141]

The new state governments behaved, in all the circumstances, with energy and sense. But there was one exception. They made it plain that blacks would not be treated as equal citizens—would, in fact, be graded as peons, as in some Latin American countries. They had freedom under the state constitutions, and provisions were made for them to sue and be sued, and to bear testimony in suits where a black was a party. But intermarriage with whites was banned by law, and a long series of special offenses were made applicable only to blacks. A list of laws governing vagrancy was designed to force blacks into semi-servile work, often with their old masters. Other provisions in effect limited blacks to agricultural labor. These Black Codes varied from state to state and some were more severe than others; but all had the consequence of relegating blacks to second-class citizenship. Plantation owners were anxious to get blacks to work as peons. Local black leaders encouraged them to sell their labor for what it would fetch, and so make freedom work. This feeling was encouraged by a new kind of federal institution, called the Freedman's Bureau, set up under the aegis of the military, which spent a great deal of bureaucratic time, and immense sums of money, on protecting, helping, and even feeding the blacks. It was America's first taste of the welfare state, even before it was established by its European progenitor, Bismarck's Germany. The Bureau adumbrated the countless U.S. federal agencies which were to engage in social engineering for the population as a whole, from the time of F. D. Roosevelt until this day. It functioned after a fashion, but it did not encourage blacks to fend for themselves, and one of the objects of the Black Codes was to supply the incentives to work which were missing.[142]

All this caused fury among the Northern abolitionist classes and their representatives in Congress. They were genuinely angry that the Southern blacks were not getting a square deal at last, and more synthetically so that the Southern whites were not being sufficiently punished. Most Northerners had no idea how much the South had suffered already; otherwise they might have been more merciful. Congress had already passed a vengeful Reconstruction Bill in 1864, but Lincoln had refused to sign it. When Congress finally reassembled in December 1865, it was apparent that this spirit of revenge was dominant, with Sumner and Stevens whipping it up, assisted by most of the Republican majority. It was clear that the president had the backing only of the small minority of Democrats. The majority promptly excluded all senators and representatives from the South, however elected, appointed a joint committee to "investigate conditions" in the "insurrectionary states," and passed a law extending the mandate of the Freedman's Bureau. Johnson promptly vetoed this last measure, lost his temper, and denounced leading Republican members of Congress, by name, as traitors. When Congress retaliated by passing a Civil Rights Bill, intended to destroy much of the

Black Codes, especially their vagrancy laws, Johnson vetoed that too. Congress immediately passed it again by a two-thirds majority, the first time in American history that a presidential veto had been overriden on a measure of importance. Thus the breach between the White House and Congress was complete. As Johnson had never been elected anyway, and had no personal mandate, his moral authority, especially in the North, was weak, and Congress attempted to make itself the real ruler of the country, rather as it was do again in the 1970s, after the Watergate scandal.[143]

The consequence was an unmitigated disaster for the South, in which the blacks ultimately became even greater victims than the whites. By June 1866, the Joint Committee reported on the South, It said that the Johnson state governments were illegal and that Congress alone had the power to reconstruct what it called the "rebel communities." It said that the South was "in anarchy," controlled by "unrepentant and unpardoned rebels, glorying in the crime which they had committed." It referred to the Fourteenth Amendment, already described, and insisted that no state government be accorded recognition, or its senators and representatives admitted to Congress, until it had ratified it. All this became the issue in the autumn 1866 midterm elections. Johnson campaigned against it, but the vulgarity and abusive language of his speeches alienated many, and he succeeded in presenting himself as more extreme, in his horrible way, than his opponents. So the radical Republicans won, and secured a two-thirds majority in both Houses, thus giving themselves the power to override any veto on their legislation which Johnson might impose. The radicals were thus in power, in a sense, and could do as they wished by law. In view of this, the governments of the Southern states would have been prudent to ratify the Fourteenth Amendment. But, as usual, they responded to Northern extremism by extremism of their own, and all but one, Tennessee, refused.

To break this impasse, the dominant Northern Radicals now attacked, with the only weapon at their disposal, the law. In effect, they began a second Reconstruction. Their object was partly altruistic—to give justice to the blacks of the South by insuring they got the vote— and partly self-serving, by insuring that blacks cast their new votes in favor of Republicans, thus making their party dominant in the South too. As it happened, most Republicans in the North did not want the blacks to get the vote. Propositions to confer it in the North were rejected, 1865–67, in Connecticut, Minnesota, Wisconsin, Ohio, and Kansas, all strong Republican states. But the Republican majority insisted nonetheless on forcing black voters on the South. In March–July 1867 it pushed through Congress, overriding Johnson's veto, a series of Reconstruction Acts, placing what they called the "Rebel States" under military government, imposing rigid oaths which excluded many whites from electoral rolls while insuring all blacks were registered, and imposing a number of conditions in addition to ratification of the Fourteenth Amendment, before any "Rebel State" could be readmitted to full membership of the Union. It also made a frontal assault on the powers of the executive branch, in particular removing its power to summon or not to summon Congress, to dismiss officials (the Tenure of Office Act) and to give orders, as commander-in-chief, to the army. Fearing obstruction by the Supreme Court, it passed a further act abolishing its jurisdiction in cases involving the Reconstruction Acts. Much of this legislation was plainly unconstitutional, but Congress planned to make it efficacious before the Court could invalidate it.[144]

This program, characteristic of the tradition of American fundamentalist idealism at its most extreme and impractical, had some unfortunate consequences. In Washington itself it led to a degree of bitterness and political savagery which was unprecedented in the history of the republic. In the debates of the 1840s and 1850s, Calhoun, Webster, Clay, and their colleagues, however much they might disagree even on fundamentals, had conducted their arguments within a framework of civilized discourse and with respect for the Constitution, albeit they interpreted it in different ways. And, in those days, Congress as a whole had treated the other branches of government with courtesy, until the Rebellion, by refusing to accept the electoral verdict of 1860, ruined all. Now the Republican extremists were following in the footsteps of the secessionists, and making a harmonious and balanced government, as designed by the Founding Fathers, impossible.

The political hatred which poisoned Washington life in 1866–67 exceeded anything felt during the Civil War, and it culminated in a venomous attempt to impeach the president himself. Johnson regarded the Tenure of Office Act as unconstitutional, and decided to ignore it by sacking Stanton, the war secretary. Stanton had always been an unbalanced figure, politically, whom Lincoln had brought in to run the War Department simply because of his undoubted energy, drive, and competence. But with the peace Stanton became increasingly extreme in using military power to bully the South. He also, like the president, had an ungovernable temper and lost it often. Johnson saw him as the Trojan Horse of the Radical Republicans within his own Cabinet, and kicked him out with relish. The Republican majority retaliated by impeaching him, under Article I, Sections 1, 2, and 5, of the Constitution. Article II, Section 4, defines as impeachment offenses "Treason, Bribery or other High Crimes and Misdemeanors." This last phrase is vague. One school of thought argues it cannot include offenses which are not indictable under state or federal law. Others argue that such non-indictable offenses are precisely what an impeachment is for— political crimes against the Constitution which no ordinary statute can easily define.

The procedure for impeachment is that the House presents and passes an impeachment resolution and the Senate convicts, or not, by a two-thirds vote. Since 1789, the House has successfully impeached fifteen officials, and the Senate has removed seven of them, all federal judges.[145] Johnson was the first president to be impeached, and the experience was not edifying. Johnson was subjected during the proceedings to torrents of personal abuse, including an accusation that he was planning to use the War Department as a platform for a personal *coup d'état,* and much other nonsense. An eleven-part impeachment resolution passed the House on February 24, 1868. There was then a three-month trial in the Senate, at the end of which he was acquitted (May 26, 1868) by 35 to 19 votes, the two-thirds majority not having been obtained. No constructive purpose was served by this vendetta, and the only political consequence was the discrediting of those who conducted it.[146]

The consequences for the South were equally destructive. The Acts of March 1867 led to a new Reconstruction along Republican, anti-white lines. Registration was followed by votes calling conventions, and these by the election of conventions, the drafting of constitutions, and their approval by popular vote. But those who took part in this process were blacks, guided by Northern army officers, a few Northerners, and some renegade whites. This new electorate was organized by pressure groups called Union Leagues, which built up a Republican Party of the South. In fact, the state constitutional conventions were almost identical with Republican nominating conventions. The new party and the imposed state were one. It was as though the

North, with its military power, had imposed one-party dictatorships on all the Southern states. The vast majority of whites boycotted or bitterly opposed these undemocratic procedures. But for the time being there was nothing they could do. Only in Mississippi did they succeed in rejecting the new constitution.

By the summer of 1868 all Southern states except three (Texas, Mississippi, and Virginia) had gone through this second, Congressional-imposed Reconstruction, and by an Omnibus Act seven of them were restored to Congressional participation (Alabama had already passed the test). As a result of the disenfranchisement of a large percentage of Southern white voters, and the addition of black ones, organized as Republicans, the ruling party carried the elections of 1868. General Grant, who had been nominated unanimously by the Republican Convention as candidate, won the electoral college by 214 votes to 80 for the Democrat, Governor Horatio Seymour of New York (1810–86). Without the second Reconstruction, it is likely Grant would have lost, and some of the Republicans, such as Sumner and Stevens, admitted that Congress had recognized the eight Southern states in 1868 primarily to secure their electoral votes. Thus America, after abolishing the organic sin of slavery, witnessed the birth of an organic corruption in its executive and Congress.[147]

These transactions at least had the merit of enabling Congress to bully the South into ratifying the Fifteenth Amendment, which stated that the right of American citizens to vote should not be denied or abridged "on account of race, color or previous condition of servitude." On the other hand, in evading its implications, Southerners could later cite, as moral justification, the fact that they had ratified it only under duress—especially true in Georgia, for instance, which had to be placed yet again under military occupation and Reconstructed for the third time. Moreover, the Republican-imposed governments in the Southern states, as might have been expected, proved hopelessly inefficient and degradingly corrupt from the start. The blacks formed the majority of the voters, and in theory occupied most of the key offices. But the real power was in the hands of Northern "carpetbaggers" and a few Southern white renegades termed "scalawags." Many of the black officeholders were illiterate. Most of the whites were scoundrels, though there were also, oddly enough, a few men of outstanding integrity, who did their best to provide honest government. There were middle-class idealists, often teachers, lawyers or newspapermen who, as recent research now acknowledges, were impelled by high motives. But they were submerged in a sea of corruption. State bonds were issued to aid railroads which were never built. Salaries of officeholders were doubled and trebled. New state jobs were created for relatives and friends. In South Carolina, where the prescriptions had been particularly savage, and carpetbaggers, scalawags, and blacks had unfettered power, both members of the legislature and state officials simply plunged their hands into the public treasury. No legislation could be passed without bribes, and no verdicts in the courts obtained without money being passed to the judges. Republicans accused of blatant corruption were blatantly acquitted by the courts or, in the unlikely event of being convicted, immediately pardoned by the governor.[148]

The South, its whites virtually united in hatred of their governments, hit back by force. The years 1866–71 saw the birth of the Ku Klux Klan, a secret society of vigilantes, who wore white robes to conceal their identities, and who rode by night to do justice. They were dressed to terrify the black community, and did so; and where terror failed they used the whip and the noose. And they murdered carpetbaggers too. They also organized race-riots and racial lynchings. They were particularly active at election-time in the autumn, so that each contest was marked by violence and often by murder. Before the Civil War, Southern whites had despised the blacks and occasionally feared them; now they learned to hate them, and the hate was reciprocated. A different kind of society came into being, based on racial hatred. The Republican governors used state power in defense of blacks, scalawags, and carpetbaggers, and when state power proved inadequate, appealed to Congress and the White House. So Congress conducted inquiries and held hearings, and occasionally the White House sent troops. But the blacks and their white allies proved incapable of defending themselves, either by political cunning or by force. So gradually numbers prevailed. The whites, after all, were in a majority, and America, after all, was a democracy, even in the South. Congressional Reconstruction gradually crumbled. The Democrats slowly climbed back into power. Tennessee fell to them in 1869, West Virginia, Missouri, and North Carolina in 1870, Georgia in 1871, Alabama, Texas, and Arkansas in 1874, Mississippi in 1875. Florida, Louisiana, and South Carolina were held in the Republican camp only by military force. But the moment the troops were withdrawn, in 1877, the Republican governments collapsed and the whites took over again.

In short, within a decade of its establishment, Congressional Reconstruction had been destroyed. New constitutions were enacted, debts repudiated, the administrations purged, cut down, and reformed, and taxation reduced to prewar levels. Then the new white regimes set about legislating the blacks into a lowly place in the scheme of things, while the rest of the country, having had quite enough of the South, and its blacks too, turned its attention to other things. Thus the great Civil War, the central event of American history, having removed the evil of slavery, gave birth to a new South in which whites were first-class citizens and blacks citizens in name only. And a great silence descended for many decades. America as a whole did not care; it was already engaged in the most astonishing economic expansion in human history, which was to last, with one or two brief interruptions—and a world war—until the end of the 1920s.

2

Two Kinds of Nobility:
Lincoln and Lee

Abraham Lincoln (1809–1865) comes high on the list of enduring popular heroes, at any rate in the English-speaking world. One crude but useful test is the number of times a person, real or imaginary, has been featured in movies. On a list compiled at the end of the twentieth century, Lincoln appeared fifth, with 137 entries devoted to him. The four ahead of him were Sherlock Holmes (211), Napoleon (194), Dracula (161) and Frankenstein (159). Lincoln thus did better than any other real figure except Napoleon. By comparison other real American heroes were far behind: Ulysses S. Grant (50), the successful general Lincoln found at long last, followed by Washington (38) and the "heroic booby," Custer (33).

There is, I think, one word that explains Lincoln's heroic preeminence in the hearts and minds of so many: goodness. He was a good man on a giant scale, a man who raised goodness into a political principle, into a way of public life, and into a code of government activity. And the fact that he came from nothing and nowhere, had little formal education, Christian training or parental guidance—had taught himself morality and made himself a good man entirely by the intelligent cultivation of sound, deep-rooted instincts—makes his character all the more appealing. There is a famous photograph of him taken at the height of the Civil War,* when things were going badly for the North. In an attempt to stir up the extraordinarily supine, not to say pusillanimous, General George B. McClellan, Lincoln visited the headquarters of the Army of the Potomac, and was snapped with the entire staff.

These officers were mostly tall for their times, but Lincoln towers over them to a striking degree. It was as though he was a different order of humanity, not a member of a master race but a higher one. And so in a sense he was. There were great men in Lincoln's day—Gladstone, Disraeli, Tolstoy, Dickens, Bismarck, Ruskin and Newman, for example. And in America, major spirits like Sherman, Grant, Whitman abounded. Yet Lincoln seems to me to have been of a different order of moral magnitude, and indeed of intellectual heroism.

Unlike all the others I have mentioned, he appeared to have no real weakness, and in scruti-

---

* [Editor's note: In the next few pages, information repeats from the previous chapter but remains necessary to the present text.]

nizing his record it is impossible to point to particular episodes and say "Here he was morally wrong," "There he was inexcusably weak," or "In this case he demeaned himself." Was he, as millions of Americans believed, sent by God— or, as angry men and women in the South were convinced, an emissary of Satan? He seemed at the time, and still seems, somehow superhuman.

All this despite the fact that his life is remarkably well documented and all the evidence has been sifted over and over again. The Lincoln Papers in the Library of Congress alone fill 97 reels of microfilm, and the library also has the Herndon-Weik collection of supporting documents on 15 reels. The big Lincoln bibliography, now over sixty years old, lists 3,958 books on Lincoln, and thousands more have appeared since. Scores of these books are of high quality and great length, and any biographer of the man must, or ought to, read them all.

Yet it is not necessary to delve deep into the documentation and the literature to grasp the essence of the man. Against any episode of the historical background in which he figured, he is both salient and transparent. One reason is physical. Like Washington he towered above the rest. But he also had a striking head and face of a rugged and lined, almost ugly, nobility. That head said it all. Or, rather, it did not say it all, for what needed to be said, Lincoln said or wrote with sublime power. If ever a statesman was a master of words, he was. Perhaps the fact that he was largely self-educated brought him to words with a freshness and sense of discovery so easily lost in the academic pursuit of literary excellence. But there is nothing naive or primitive about Lincoln's use of English. It is simple; but also extremely sophisticated. He chose words not for their grace or glory but for their fundamental accuracy and truthfulness. And in this pursuit of truth he achieved grace and glory as well.

Here is a simple example. Lincoln rose from nothing to the White House through the law. From being a manual laborer in a variety of humble occupations—rail-splitting was the most skilled—he acquired enough book knowledge to set up as a lawyer. He did sufficiently well in his profession to be able to do what it taught him was an overwhelming necessity: to change bad laws by political action. People often say America has too many lawyers. It is even asserted, perhaps justly, that there are more lawyers in America than in the rest of the world put together. That is because it is easier to become a lawyer in America than in any other country, and always has been. If Lincoln had been born in England, France or Germany, it is most unlikely he could have become a lawyer, and we would never have heard of him. As it was, he got ahead. But he was a lawyer with a difference. He became a skilled lawyer but remained a good man. Here is a letter which sums him up:

Springfield, Illinois
21 February 1856
To Mr. George P. Floyd,

Quincy, Illinois

    Dear Sir,

        I have just received yours of 16th, with check on Flagg & Savage for twenty-five dollars. You must think I am a high-priced man. You are too liberal with your money.

        Fifteen dollars is enough for the job. I send you a receipt for fifteen dollars, and return to you a ten-dollar bill.

Yours truly,
A. Lincoln

I would like this letter framed and hung on the partners' desks of every law firm in the country. It is brief, simple and embodies action—the enclosed ten-dollar bill—rather than verbal waffle. Not that Lincoln was a softy. In another case, where he had greatly benefited the Illinois Central Railroad Company by a successful piece of law work, they refused to pay what he regarded as a fair fee. He sued them, and got it.

Mr. Floyd, the recipient of the letter, must surely have felt, on reading it, that the writer was a good man. So must Michael Hahn, the recipient of the next letter. The date is March 13, 1864, and Lincoln was writing from the Executive Mansion, as the White House was then called. Louisiana had already been occupied by the North, and Hahn had been installed as governor. Lincoln wanted Hahn to give some blacks the vote quickly, but the letter shows him as a practical statesman, an empiricist, seeking to do good by stealth.

Private
The Hon. Michael Hahn

My dear Sir,
I congratulate you on having fixed your name in history as the first free-state Governor of Louisiana. Now you are about to have a convention which, among other things, will probably define the elective franchise. I barely suggest for your private consideration, whether some of the colored people may not be let in—as, for instance, the very intelligent and especially those who have fought gallantly in our ranks. They would probably help, in some trying time to come, to keep the jewel of liberty within the family of freedom. But this is only a suggestion, not to the public, but to you alone.

Yours truly,
A. Lincoln

I quote this letter because it struck me, when I saw it in facsimile in an auction catalog (1994), as another example of the way Lincoln's mind worked, and because I can't remember ever having seen it in print. It shows the suggestive, subtle, intuitive side of Lincoln very well, and is characteristically lit up by a striking phrase, "to keep the jewel of liberty within the family of freedom," which—to judge by the appearance of the holograph—occurred to Lincoln as he was writing the letter.

Words, and the ability to weave them into webs which cling to the memory, are extremely important in forwarding political action. This was already true in semiliterate fifth-century B.C. Athens, as Thucydides makes clear, and in republican Rome, as Shakespeare, with his uncanny gift for getting history right, shows brilliantly in *Julius Caesar*. It was even more important in the third quarter of the nineteenth century in America, where most of the population was aggressively literate, and brought up to read and relish key documents—the Declaration of Independence, the Constitution, and the Bill of Rights. That was the way Lincoln himself was brought up (or brought himself up) and he added to the canon two of the first of its documents: the Gettysburg Address and the Second Inaugural Address. Entire books have been written on these speeches, and their evolution. But it seems to me that the key phrases within them came to Lincoln in intuitive flashes, leaping up from a mind that had brooded so long on the nature of political truth and justice, and the frailty of man in promoting them, that it was composed of hot coals from which sparks might be emitted at any instant.

Not that Lincoln was a furnace of rhetoric, or a Man of Destiny or a superhuman force in any way whatsoever. He is not at all the kind of person Carlyle describes in *Heroes and Hero-Worship*, and I have not so far found any reference to him in Carlyle's voluminous correspondence indicating approval. Lincoln was the first to admit that he often, and on the most important occasions, reacted to events rather than directed them. Lincoln was not a will-to-power man but a democrat. He sought to serve the republic, not to impose his ideas upon it. He was a typical American in that he believed passionately in justice, and its embodiment in the rule of law, and in the country's long-term ability always to realize these beliefs in practice. As he put it during one of his debates with Senator Douglas in 1858: "The cause of Civil Liberty must not be surrendered at the end of *one,* or even one *hundred* defeats." But he was an untypical American, and an untypical hero, in that there was a streak of melancholy in his character. He suffered from occasional bouts of depression, and some historians have even argued that he was a lifelong depressive. In 1998, for instance, Andrew Delbanco, in a series of lectures at Harvard, argued that Lincoln's private despair was the engine of his public work: "The lesson of Lincoln's life is that a passion to secure justice can be a remedy for melancholy." In another interpretation, Joshua Wolf Shenk argued in 2005 *(Lincoln's Melancholy: How Depression Challenged a President and Fueled His Greatness)* that it was precisely the depressive nature of Lincoln's mind that gave him a passion for justice in the first place. One needs to take aboard such arguments in considering such a complex man as Lincoln, but it is important not to exaggerate the role that depression played in his public activities. Lincoln was always busy, and almost always busy doing things he wanted to do and which were worth doing. He had time for thought, and no one thought harder than he did, but no time for brooding.

One way in which he was untypical of most Americans was that he did not, strictly speaking, believe in God or, at any rate, a God most of his fellow citizens would have recognized. But he certainly felt there was a guiding providence and that, in the providential scheme, Americans—"the almost-chosen people," as he called them—had an important role to play. America was a pilot state for a better world and if America failed its great test over slavery, the outlook was grim. In order to survive and lead the world, it must remain united: hence the Union was the one clear and unassailable principle in Lincoln's worldview, the one point on which he never had any hesitation or doubt.

By comparison slavery was a mere phenomenon. Lincoln thought it an evil—who in his heart did not?—but he refused to see it in inflammatory moral terms. He went out of his way to admit that Southerners were "no more responsible for the origins of slavery than we." He was prepared to live with slavery, at any rate for a time. What he was not prepared to do was see it extended, and that really was the issue on which the Civil War was fought.

Lincoln did not regard blacks as equals. Or rather, they might be morally equal but in other respects they were fundamentally different, and unacceptable as fellow citizens without qualification. He said bluntly that it was impossible just to free the slaves and make them "politically and socially our equals." He freely admitted an attitude to blacks which would now be classified as racist: "My own feelings will not admit [of equality]." The same was true, he added, of a majority of whites, North and South. "Whether this feeling accords with justice and sound judgment is not the sole question. A universal feeling, whether well- or ill-founded, can not be safely disregarded." It is such statements, and many others of a similar nature, which make Lincoln's speeches and writings so riveting. They show that his salient characteristic was candor, a willingness to admit and articulate truth, however inconvenient or unheroic or distasteful or inconsistent it might be.

Lincoln was a pragmatist as well as a democrat. He realized that if the Union were to be preserved he must carry a majority, and if possible a big majority, of the people with him. That meant he had to take account of their real feelings. They could be guided and led, up to a point. But they could not be hustled, let alone forced. Fortunately there was not an atom of fanaticism in Lincoln.

The steps by which Lincoln reached his famous decision to emancipate the slaves show his pragmatism and sense of timing at their best. No man was ever more a practical statesman, as opposed to an ideological one. Every aspiring politician, American, British or foreign, should study his career and the way he applied his mind to the fearful problems which confronted him. Lincoln was a strong man and, like most men quietly confident of their strength, without vanity or self-consciousness. There was a little incident toward the end of his life which, to me, is full of meaning. After the fall of Richmond, the Confederate capital, and on the same day Robert E. Lee finally surrendered, Lincoln went to see his secretary of state, with whom he often disagreed, and whom he did not particularly like. Seward had somehow contrived to break both his arm and his jaw. Lincoln found him not only bedridden but unable to move his head. Without a moment's hesitation, the president stretched out at full length on the bed and, resting on his elbow, brought his face near Seward's, and they held an urgent, whispered conversation on the next steps the administration should take. Then Lincoln talked quietly to the agonized man until he drifted off to sleep.

Lincoln could easily have used the excuse of Seward's incapacity to avoid consulting him at all. But that was not his way. He invariably did the right thing, however easily it might have been avoided. Of how many other great men might this be said?

At the same time, in pursuing his overwhelming objective of preserving the Union, to him a moral as well as a political necessity, Lincoln showed himself capable of great ruthlessness. He used the power of the presidency to its utmost. Here, Washington, during the Whiskey Rebellion, had set a precedent of a kind, but Lincoln's exploitation of his office, and of the Constitution, was of an altogether different order. He showed that a great republican democracy, once roused to pursue a mighty and righteous object, was capable of a forcefulness, even ferocity, which was both terrifying as well as sublime. This heroic fortitude in enlisting all the power of the Union in the cause of right itself transformed the presidency and the nation, and made it possible for Lincoln's strong-minded successors to follow the precedent on the world scene. Lincoln's ruthlessness was the guide for Woodrow Wilson in taking the United States into the First World War and making the peace that followed, for Franklin Delano Roosevelt in fighting the Second World War on the largest possible scale, for Harry S. Truman in using the atomic bomb against Japan, and in mobilizing the free world against Soviet and Communist aggression, for Ronald Reagan in winning and ending the Cold War and destroying the Soviet empire, and for George W. Bush in fighting international terrorism in its homelands.

Lincoln was able to inaugurate this new kind of heroic leadership in American history because he was a new kind of American—someone for whom citizenship of the Union was far more important than his provenance from a particular state. In the tremendous events of the Civil War, the central event in American history, Lincoln was not the only hero. The South had to have a hero too. That part could not be played by the president of the Confederacy, Jefferson Davis. He was not a negligible figure. He was in many ways virtuous, consistent, truthful, courageous and always anxious to be just. But he was also narrow and narrow-minded, extraordinarily constricted by his environment and upbringing, no more heroic than a severely blinkered cart horse painfully pulling a heavy wagon on a preordained track to nowhere.

The South, however, found a hero in Robert E. Lee. He was a noble and virtuous man, like Lincoln.

But the contrast in their motivations was significant. The two men had quite different ideas about the individual states, which had nothing directly to do with the North-South divide. Lincoln was born in Kentucky, which in the seventeenth century was quite inaccessible beyond the Alleghenies, and was not open to colonists until 1774. It was the post-1800 beneficiary of the Wilderness Road, and the "dark and bloody ground" of Indian warfare. In 1792 it was admitted to the Union as the fifteenth state, the first from beyond the mountains, and then only after Virginia ceded title to its theoretical western lands.

After 1840, using the great Ohio River from Louisville, Kentucky became a slave market to the South. It had its own Civil War: 30,000 men from Kentucky fought for the Confederacy, against 60,000 for the Union. Lincoln felt no allegiance at all to the state. When he was nineteen, his family moved to Illinois. It had been under French rule until 1763. Then it became part of the Indiana Territory. It was changed to the Illinois Territory in 1809, and was admitted as the twenty-first state in 1818. It had superb agricultural land and was potentially rich in other ways, but it did not attract attention till the Lincoln-Douglas debates of 1858, which first gave it political importance. Lincoln made Springfield his home and Illinois gave him a professional and political career. But it was to the Union, "great and strong," to which he felt allegiance and duty, as well as emotional attachment. States' rights were a fact of life of which, as a pragmatist, he took account. But they meant nothing to him spiritually. To Lee it was profoundly different. Virginia really went back to Ralegh's Roanoke colony of 1584, for when a permanent English settlement was established in Jamestown in 1607, Virginia, after Ralegh patroness Queen Elizabeth, was the automatic choice of name. For quite a long time, Virginia *was* the English presence in America, constituting all the land not occupied by Spain and the French. In 1619 the House of Burgesses was founded in Jamestown, the first representative institution set up in the New World, indeed anywhere outside Europe. There are references from this time to "the Colony and Dominion of Virginia"—hence the term "the Old Dominion" applied to it, making it different from all the others, and special. By 1624 it was a royal colony, and by 1641 it was by far the most important, with 7,500 citizens and over 1,000 prosperous farmers and plantations. Under the Commonwealth it was virtually independent and always felt itself to be, and largely was, self-governing. It was, in the 1770s, the natural leader of the rebellion, along with Massachusetts. Virginia's Peyton Randolph was elected president of the First Continental Congress in 1774. Many of the key figures in the creation of the Union were Virginian: not only Washington but also Patrick Henry, Edmund Randolph and John Marshall, the man who effectually created the Supreme Court. Seven out of the first ten presidents were Virginians.

When Lincoln was elected president and the lower South, led by the extremists of South Carolina, seceded, it was by no means clear that Virginia would follow suit. And if Virginia had stuck by the Union, the secession would have become insignificant. Many professional soldiers from Virginia, such as General Winfield Scott and George H. Thomas, made it clear they would remain Unionists whatever the Old Dominion decided. It was thought that Lee would take a similar view and Lincoln offered him the command of the new Union army that had to be created. But uncertain what Virginia would do, and determined to follow her for good or ill, Lee declined the appointment. And when in April 1861, by a democratic decision of the whites, Virginia opted for secession, Lee reluctantly went to war on her behalf. As he put it: "I prize the Union very highly and know of no personal sacrifice I would not make to preserve it, save that of honor."*

What did he mean by that? Honor was the key word in Lee's life and vocabulary. It meant something very special to him. He came from the old Virginia aristocracy and married into it. His father was Henry Lee III, revolutionary war general, congressman and one-time governor of Virginia. His wife, Ann Carter, was the great-granddaughter of Robert "King" Carter, who owned 300,000 acres and 1,000 slaves. That was the grand side of Lee's background. There was also the dark side.

To put it bluntly, his father became a crook. His claim to be appointed commander-in-chief of the U.S. army was dismissed by George Washington

* [Editor's note: In the next few pages, information repeats from the previous chapter but remains necessary to the present text.]

with the euphemistic "lacks economy." He was certainly a big spender, and to finance his tastes he
became a dishonest land speculator. Among those
he defrauded was Washington himself. He was
given the ironic nickname "Light Horse Harry,"
and eventually went bankrupt, and was jailed twice.
When Robert was six, his father fled from his credi-
tors to the Caribbean, and never returned. His
mother was left a needy widow with many children.
The family's reputation was not improved by a ruffianly stepson known as "Black Horse Harry," who
specialized in adultery.

Robert E. Lee seems to have set himself up, quite
deliberately, to redeem the family honor by leading an exemplary life of public service. "Honor," a
word he pronounced with a special loving empha-
sis, putting a stress on each syllable, meant everything to him. His dedication to honor made him
a peculiarly suitable person to become the equivalent to the South of Lincoln, sanctifying its cause
by personal probity and virtuous inspiration. Like
Lincoln, though in a less egregious and angular
manner, he looked the part. He radiated beauty
and grace. He sat his famous warhorse, Traveller,
in a statuesquely erect and distinguished posture,
the fine stallion too looking the part. Though he
was almost six foot, he had small hands and feet,
and there was something feminine in his sweetness
and benignity. His fellow cadets at West Point called
him "The Marble Model." With his fine beard, first
tinged with gray, then white, he became in his fifties
a Homeric patriarch. Photos of him remind one of

the dignified heads of the Roman emperors around the Sheldonian Theatre in Oxford. It is surprising to learn that he was just sixty-three when he died, loaded with tragic honors.

After an industrious youth, he led a blameless life at West Point, and actually saved from his meager pay, at a time when all other Southern cadets prided themselves on acquiring debts. His high grades meant he was accepted by the elite Army Corps of Engineers, in an army whose chief occupation was building forts. His specialty was taming the wild and mighty river which Mark Twain described so unforgettably in *Life on the Mississippi*. Lee served with valor and immense success in the Mexican war of 1846–1848, emerging a full colonel. Then followed posts as superintendent at West Point and cavalry commander against the Plains Indians. In 1859 Lee put down John Brown's rebellion at Harpers Ferry, and reluctantly handed him over to be hanged. Lee owned slaves much of his life but, like most educated Virginians, thought slavery a great evil, which damaged the whites even more than the blacks. (In this he differed profoundly from Jefferson Davis, who actually believed slavery was beneficial to blacks.) Lee joined the South not to preserve slavery but to enable the Old Dominion to preserve its traditional self-government. It was a point of honor, as he saw it.

Lee cannot have been happy with the way the South ran its war. It is important to remember that, whereas Lincoln was able to run a centralized government which at moments amounted to a virtual dictatorship, the South remained a confederacy, with each state retaining elements of sovereignty, not least over its armed forces. It was also handicapped by many other burdens arising from its ideology, not least Jefferson Davis' policy of defending the frontiers of all the Confederate states, making a concentration of its limited armed forces impossible. Up to half were permanently employed on pointless frontier duties. This dispersal of effort went directly against Lee's own view of the strategy the South must pursue if it were to survive. Unlike most people, on both sides, he predicted from the start that the war would be long and bloody. But he grasped that the South had a commitment to the war which many, perhaps most, Northerners lacked. The North had much greater resources of all kinds and must win in the end, unless the South could play upon the North's relative lack of commitment. Its only chance of winning was to engage the bulk of the Union forces in a decisive battle, and win it. This would provoke a political crisis in the North, perhaps force Lincoln's resignation, and open the road to a compromise—which is what Lee had wanted all along.

Lee had an excellent command of tactics as well as a sure sense of strategy, and held high command in some of the bloodiest battles, winning Bull Run, Fredericksburg and Chancellorsville. But he lacked the supreme authority his strategy required. He was not appointed general-in-chief of the Southern forces until February 1865, far too late and only two months before he was obliged to surrender them at Appomattox. Moreover, as a commanding general he had weaknesses. He lacked the killer instinct. Watching his men delightedly chase the beaten and fleeing Unionists at Fredericksburg, he sadly remarked: "It is well that war is so terrible. Otherwise we should grow too fond of it." And he was too diffident to be a great commander. He disliked rows and personal confrontations, inevitable in war if a general is to assert his authority. He preferred to work through consensus. He tended to issue guidance to subordinate commanders rather than detailed, direct orders. At Gettysburg, the gigantic battle he had been waiting for, which gave him a real chance to destroy the main Union army, this weakness proved fatal. Lee's success on the first day was overwhelming, but on the second he did not make it clear to General James Longstreet that he wanted Culp's Hill and Cemetery Ridge taken at all costs. Longstreet provided too little artillery support to Pickett's famous charge. Even so, a few of Pickett's men reached the crest, and it would have been enough, and the battle won, if Longstreet had thrown in all his men as reinforcements. But he did not do so and the battle was lost. Lee sacrificed a third of his men and the Confederate army was never again capable of winning the war. "It has been a sad day for us," said Lee at one o'clock the following morning, "almost too tired to dismount." He added: "I never saw troops behave more magnificently than Pickett's division . . . And if they had been supported as they were supposed to have been—but for some reason, not yet fully explained to me, they were not—they would have held the position and the day would have been ours." Then he paused, and said, "in a loud voice: 'Too bad! *Too bad!* OH, TOO BAD!'"

Lee was a true hero. He insisted on making possible for others the freedom of thought and action he sought for himself. That is a noble aim, but it is not a virtue in a commanding general. "*C'est magnifique, mais ce n'est pas la guerre.*"

After the war, Lee took on the thankless task of running a poor university, Washington College. But he was broken and tired and did not last long. His life was a protracted elegy for the lost South and its noble values, which were perhaps more myth than reality but were nonetheless treasured in his heart. But there was a surprising element of laughter in all his woes. He had a sense of the absurdity of life, as well as its tragedy. When a wartime admirer in Scotland sent him a superb Afghan rug and a tea cozy, Lee delightedly draped the rug around his shoulders, donned the cozy as a helmet and did a little dance while his daughter Mildred played the piano.

He never struck heroic poses. He was modest to a fault, hid from publicity and when in doubt kept his mouth shut. Southern mythmakers have it that his famous last words recalled Gettysburg: "Tell Hill he *must* come up!" and "Strike the tent!" In fact he said nothing. Lincoln too left behind no famous last words. After such lives, what is there to say?

Source Notes

The Almost Chosen People

1. J. M. Murrin in R. Beeman *et at.* (eds): *Beyond Confederation: Origins of the Constitution and of American National Identity* (Chapel Hill 1987), 346–7.

2. Edwin Haviland Miller, *Salem is My Dwelling Place*, 379–81.

3. R. F. Nichols: *Franklin Pierce* (2nd edn New York 1958), 75.

4. Miller, *opus cit.*, 383–4.

5. There is a facsimile edition of Hawthorne's *Life of Franklin Pierce* (Boston 1970), with an introduction by R. C. Robey.

6. Nichols, *opus cit.*, 216.

7. W. C. Davis: *Jefferson Davis: the Man and His Hour* (New York 1991), 251.

8.   Robert E. May: *The Southern Dream of a Caribbean Empire* (Baton Rouge 1973), 60ff.

9.   Lawrence Greene: *Filibuster: the Career of William Walker* (New York 1937).

10. K. S. Davis: *Kansas: a Bicentennial History* (New York 1976), 47ff; R. W. Johannsen: *Stephen A. Douglas* (New York 1973).

11. P. W. Gates: *Fifty Million Acres: Conflicts Over Kansas Land Policy, 1854–90* (New York 1954).

12. J. A. Rawley: *Race and Politics: "Bleeding Kansas" and the Coming of the Civil War* (New York 1969); S. B. Oates: *To Purge This Land with Blood: John Brown* (New York 1970).

13. D. H. Donald: *Charles Sumner*, 2 vols (New York 1960–70).

14. Samuel Eliot Morison and Henry Steele Commager, *The Growth of the American Republic*, i 654ff.

15. J. T. Carpenter: *The South as a Conscious Minority* (Baton Rouge 1930); A. O. Craven: *The Growth of Southern Nationalism, 1848–1861* (New York 1953).

16. Louis Hacker: *Triumph of American Capitalism* (New York 1940), 281ff; R. B. Flanders: *Plantation Slavery in Georgia* (Atlanta 1933), 221–3; C. S. Sydnor: *Slavery in Mississippi* (New Orleans 1933), 196ff.

17. Ulrich B, Phillips: *American Negro Slavery* (New York 1918).

18. L. C. Gray: *History of Agriculture in the Southern United States to 1850*, 2 vols (Richmond 1933), i 460ff.

19. W. E. Dodd: *The Cotton Kingdom* (New York 1921), 121.

20. C. and M. Beard: *Rise of American Civilisation*, 4 vols (New York 1917–42), ii 5–6.

21. H. U, Faulkner, *American Economic History*, 320.

22. Quoted in *Niles Weekly Register*, April 19, 1845.

23. David Herbert Donald: *Lincoln* (London 1995), 19–20; for the early Lincoln see E. Hertz (ed.): *The Hidden Lincoln: from the Letters and Papers of William H. Herndon* (New York 1938); L. A. Warren: *Lincoln's Youth: Indiana Years, Seven to 21, 1816–30* (New York 1959); C. B. Strozier: *Lincoln's Quest for Union: Public and Private Meanings* (New York 1982). Beware of unreliable transcripts in Herndon papers and of psychobabble in Strozier.

24. For Lincoln details, see M. E. Neely Jr.: *The Abraham Lincoln Encyclopaedia* (New York 1982). For his suicide fears, see H. I. Kushner: *Self-Destruction in the Promised Land: a Psychocultural Biology of American Suicide* (New Brunswick 1989), Chapter 5.

25. W. C. Temple and H. E. Pratt: "Lincoln in the Black Hawk War," *Bulletin of the Abraham Lincoln Association*, 54 (December 1938), 3ff.

26. For Lincoln as a lawyer, see A. A. Woldman: *Lawyer Lincoln* (Boston 1936); J. J. Duff: *A. Lincoln: Prairie*

Lawyer (New York 1960); J. P. Frank: *Lincoln as a Lawyer* (Champaign 1961).

27. For Ann Rutledge and her effect on Lincoln, see Donald, *opus cit.*, 608 n. 55.

28. See John Lorring: *Tiffany's 150 Years* (New York 1987), 46–7; for Mrs Lincoln, see R. P. Randall: *Mary Lincoln: Biography of a Marriage* (Boston 1953) and her *The Courtship of Mr Lincoln* (Boston 1957).

29. For Lincoln's spell in Congress, see Donald, *opus cit.*, Chapter 5; D. E. Riddle: *Congressman Abraham Lincoln* (Westport 1979); Paul Findley: *Abraham Lincoln: the Crucible of Congress* (New York 1979).

30. *Sayings and Anecdotes of Lincoln* (New York 1940), 107–8.

31. Jean Baker: *Mary Todd Lincoln* (New York 1960).

32. W. J. Wolf: *The Almost Chosen People: a Study of the Religion of Abraham Lincoln* (New York 1959).

33. For Lincoln's depressions, see Donald, *opus cit.*, 163ff.

34. For the writings and speeches of Lincoln, I have used Don. E. Fehrenbacher (ed.): *Abraham Lincoln: Speeches and Writings,* 2 vols (Classics of Liberty Library, New York 1992).

35. For Lincoln on slavery, see Donald, *opus cit.*, 165–7, 180–1, etc.

36. *Ibid.*, 191; *Speeches and Writings*, 365; W. E. Gienapp: *The Origins of the Republican Party, 1852–56* (New York 1987); the supposed full text of the speech published in the September 1896 issue of *McClure's Magazine* had been questioned.

37. Roy P. Basler: *Collected Works of Abraham Lincoln*, 8 vols (New Brunswick 1953), ii 341.

38. *Herndon's Lincoln*, ii 384.

39. *Speeches and Writings*, i 426–34; Don E. Fehrenbacher: *Prelude to Greatness: Lincoln in the 1850s* (Stanford 1962); Donald, *opus cit.*, 206ff.

40. For the debates, see R. A. Heckman: *Lincoln v. Douglas: the Great Debates Campaign* (Washington, D.C. 1967); texts in R. W. Johannsen: *The Lincoln-Douglas Debates of 1858* (New York 1965).

41. Speech at Clinton, September 8, 1858.

42. See David Zarefsky: *Lincoln, Douglas and Slavery: in the Crucible of Public Debate* (Chicago 1990).

43. *Speeches and Writings*, ii 106–8; the long campaign autobiography is ii 160–7.

44. Published in the *New York Times*, January 24, 1854; for Chase,. see David Donald (ed.): *Inside Lincoln's Cabinet: the Civil War Diaries of Salmon P. Chase* (New York 1954); for Seward, see G. G. Van Deusen: *William Henry Seward* (New York 1967).

45. See the two letters, May 19 and 23, 1860, *Speeches and Writings*, ii 156–7.

46. *Ibid.*, ii 111–30.

47. Among the many recent books on anti-slavery agitation, the best are: Thomas Bender (ed.): *The Antislavery Debate: Capitalism and Abolitionism as a Problem in Historical Interpretation* (Berkeley 1992); Alan M. Kraut (ed.): *Crusaders and Compromisers: Essays on the Relationship of the Antislavery Struggle to the Antebellum Party System* (Westport 1983); L. Perry and M. Fellman (eds): *Antislavery Reconsidered: New Perspectives on the Abolitionists* (Baton Rouge 1979).

48. J. C. Furnas: *The Road to Harper's Ferry* (New York 1959).

49. Elting Morison, "The Election of 1860," in Arthur M. Schlesinger Jr. and F. R. Israel (eds), *American Presidential Elections,* ii 1097–122. See also W. E. Gienapp: "Who Voted for Lincoln?," in J. L. Thomas (ed.): *Abraham Lincoln and the American Political Tradition* (Amherst 1986), 50ff.

50. For Davis, see William C. Davis: *Jefferson Davis: the Man and His Hour* (New York 1991), esp. 689ff.

51. For Davis' background, see the life by his widow, Varina H. Davis: *Jefferson Davis,* 2 vols (Charleston 1890) and the "official" collection, *Jefferson Davis, Constitutionalist, His Letters, Papers and Speeches,* ed. Dunbar Rowland, 10 vols (Baton Rouge 1923).

52. For Davis' treatment of slaves etc., see Jefferson Davis: *Rise and Fall of the Confederate Government,* 2 vols (New York 1881), i 518; Varina Davis, *opus cit.,* i 174–9.

53. William C. Davis, *opus cit.*, 125.

54. *Ibid.*, 198–9.

55. *Ibid.*, 127–67.

56. The row is described in Winfield Scott: *Memoirs*, 2 vols (New York 1864), and in William C. Davis, *opus cit.*, 228ff.

57. T. C. Cochran: *Frontiers of Change: Early Industrialism in America* (New York 1981), 73.

58. William C. Davis, *opus cit.*, 258–60.

59. *New Orleans Bee*, December 14, 1860.

60. William C. Davis, *opus cit.*, 283.

61. For Lincoln during this vital period, see W. E. Baringer: *A House Dividing: Lincoln as President Elect* (Springfield, Illinois 1945).

62. William C. Davis, *opus cit.*, 296.

63. D. W. Meinig, *Continental America*, 477–8.

64. Morison and Commager, *opus cit.*, 667ff.

65. William C. Davis, *opus cit.*, 270.

66. For the early weeks of the Lincoln presidency, see P. S. Paludan: *The Presidency of Abraham Lincoln* (Lawrence 1994), Chapters 2–3.

67. R. N. Current: *Lincoln and the First Shot* (New York 1963).

68. Quoted in Emory M. Thomas: *Robert E. Lee: a Biography* (New York 1995), 188.

69. R. A. Wooster: *Secession Conventions of the South* (New York 1962) for details.

70. Quoted in Drew Gilpin: *James Henry Hammond* (Baton Rouge 1982).

71. See R. L. Andreano (ed.): *Economic Impact of the Civil War* (New York 1962); H. N. Scheiber: "Economic Change in the Civil War Era: Analysis of Recent Studies," *Civil War History,* 11 (1965), 396ff.

72. H. D. Capers: *Life and Times of C. G. Memminger* (New York 1893).

73. See the excellent summary of Southern finances by J. C. Schwab: "The South During the War, 1861–65," in *Cambridge Modern History* (Cambridge 1934), vii 603ff; Davis, *opus cit.*, 601ff.

74. Elizabeth Merritt: *James Henry Hammond* (Baton Rouge 1923).

75. For a summary of this incident, see C. F. Adams: "The Trent Affair," *Massachusetts Historical Society Proceedings* 45 (1911), 35ff.

76. Davis, *opus cit.*, 319.

77. *Ibid.*, 366.

78. T. L. Livermore: *Numbers and Losses in the Civil War* (New York 1901).

79. For the influence of geography on the conflict, see Meinig, *opus cit.*, 494ff.

80. W. B. Yearns; *The Confederate Congress* (New York 1960); F. L. Owsley: *States Rights in the Confederacy* (New York 1925); Davis, *opus cit.*, 444ff; R. D. Meade: *Judah P. Benjamin* (New York 1943).

81. Davis, *opus cit.*, 447.

82. Donald, *Lincoln*, 362–4; text of Lincoln's proclamation in *Speeches and Writings*, ii 318–19.

83. Quoted in Donald, *Lincoln*, 368.

84. Different texts circulated; see *Speeches and Writings*, ii 357–8; Lincoln's *Collected Works*, v 388–9.

85. See Joseph H. Parks: *General Leonidas Polk, CSA: Fighting Bishop* (New York 1962).

86. J. W. Silver: *Confederate Morale and Church Propaganda* (New Orleans 1957).

87. See the special issue, "Civil War Religion," *Civil War History*, 6 (1960).

88. See J. G. Randall and R. N. Current: *Lincoln the President: Last Full Measure* (New York 1955), chapter entitled "God's Man."

89. Chester F. Dunham: *Attitude of the Northern Clergy Towards the South*, 1860–65 (New York 1942); D. W.

Harrison: "Southern Protestantism and Army Missions in the Confederacy," *Mississippi Quarterly*, 17 (1965), 179ff.

90. W. J. Wolf: *The Almost Chosen People: a Study of the Religion of Abraham Lincoln* (New York 1959).

91. J. H. Franklin: *The Emancipation Proclamation* (New York 1963); Donald, *Lincoln*, 366ff.

92. *Cambridge Modern History*, vii, Chapter xviii, "The North During the War: Finance"; B. W. Rein: *Analysis and Critique of Union Financing of the Civil War* (New York 1962); Bray Hammond: *Sovereignty and the Empty Purse: Banks and Politics in the Civil War* (New York 1970); A. M. Davis: *Origins of the National Banking System* (New York 1910).

93. B. P. Thomas and H. M. Hymam: *Stanton* (New York 1962); W. W. Hassler: *General George B. McClellan* (New York 1957).

94. See Allan Pinkerton's autobiography: *Criminal Reminiscences and Detective Sketches* (New York 1879) and *Thirty Years a Detective* (Chicago 1884).

95. For Jackson, see G.F.R. Henderson: *Stonewall Jackson and the American Civil War*, 2 vols (New York 1898); there are many modern books dealing with him.

96. Mrs. James Chesnut: *A Diary from Dixie* (New York 1949); General Richard Taylor: *Destruction and Reconstruction: Personal Reminiscences of the Late War* (New York 1879); Edmund Wilson: *Patriotic Gore* (New York 1962), 279ff, 303–4.

97. Wilson, *opus cit.*, 300.

98. John Esten Cooke: *Wearing the Grey* (New York 1867).

99. *Ibid.*

100. Lee is entombed in Douglas Southall Freeman: *R. E. Lee: a Biography*, 4 vols (New York 1934–5); the best life is E. M. Thomas: *Robert E. Lee: a Biography* (New York 1995).

101. Thomas, *opus cit.*, 187ff.

102. See E. B. Coddington: *The Gettysburg Campaign: a Study in Command* (New York 1968). See also J. Luvass and H. W. Nelson (eds): *The US Army War College Guide to the Battle of Gettysburg* (Carlisle 1986); Thomas, *opus cit.*, 287ff.

103. E. S. Miers: *Web of Victory: General Grant at Vicksburg* (New York 1955).

104. Grant wrote one of the great American autobiographies, *Personal Memoirs*, 2 vols (New York 1885–6).

105. For Lincoln's relations with Grant, see T. H. Williams: *Lincoln and His Generals* (New York 1952).

106. For these battles and casualties, see R. U. Johnson and C. C. Buel (eds): *Battles and Leaders of the Civil War* (New York 1884–8), iv; for Lincoln's sorrow, Donald, *Lincoln*, 500.

107. Davis, *opus cit.*, 531, 544, 594.

108. Quoted in Wilson, *opus cit.*, 271.

109. J. M. Gibson: *Those 163 Days: Sherman's March* (New York 1961).

110. Bruce Catton: *Grant Takes Command* (New York 1969).

111. H. M. Hyman: "The Election of 1864," in Schlesinger and Israel, *opus cit.*, ii; E. C. Kirkland: *Peacemakers of 1864* (New York 1927).

112. Text in *Speeches and Writings*, 686–7.

113. Miller, *opus cit.*, 474.

114. Justin Kaplan: *Walt Whitman: a Life* (New York 1980), 260–1.

115. Donald, *Lincoln*, 580–1. This was April 9, 1865, the date of Lee's surrender.

116. *Macbeth*, Act II, Scene 2. Adolphe de Chambrun: *Impressions of Lincoln and the Civil War: a Foreigner's Account* (New York 1952), 82.

117. Kaplan, *opus cit.*, 262, 297.

118. Stewart Brooks: *Civil War Medicine* (Springfield, Illinois 1966), 97; R. M. Buck: *Walt Whitman* (Philadelphia 1883), 37, quoted in Kaplan, *opus cit.*, 266n.

119. For extracts from Alcott's letters, see Edna D. Cheney: *Louisa May Alcott: Her Life, Letters and Journals* (New York 1889).

120. Judith Farr: *The Passion of Emily Dickinson* (Cambridge 1992), for a recent view of all this.

121. John D. Unruh Jr.: *The Plains Across: the Overland Emigrants and the Trans-Mississippi West, 1840–1860* (Urbana 1979).

122. For such incidents, see Francis Paul Prucha: *The Great Father: the United States Government and the American Indians*, 2 vols (Lincoln, Nebraska 1984) and the trilogy by Robert M. Utley: *Frontier Regulars: the U.S. Army and the Indian, 1866–91* (New York 1973); *The Indian Frontier of the American West 1846–90* (Albuquerque 1984), and *The Last Days of the Sioux Nation* (New Haven 1963).

123. See P. W. Gates and R. W. Swenson: *History of Public Land Law Development* (New York 1968).

124. See E. S. Pomeroy: *The Territories and the United States, 1861–90* (Philadelphia 1947) and J. E. Eblen: *The First and Second United States Empires: Governors and Territorial Governments, 1784–1912* (Pittsburgh 1968).

125. Mark Twain: *Roughing It*; see also C. A. Milner *et al.* (eds): *The Oxford History of the American West* (New York 1994), 201ff.

126. F. L. Paxson: *The Last American Frontier* (New York 1910), 170ff.

127. Davis, *opus cit.*, 598ff.

128. *Rise and Fall of the South*, ii 518ff.

129. Thomas, *opus cit.*, 361ff.

130. For these events, see B. H. Liddell Hart (ed.): *Sherman's Memoirs*, 2 vols (New York 1957).

131. For the background to the conspiracy to murder Lincoln, see W. A. Tidwell *et al.*: *The Confederate Secret Service and the Assassination of Lincoln* (Jackson 1988); for the event itself, see W. E. Reck: *Abraham Lincoln: His Last 24 Hours* (Jefferson 1987). The Surratt Society has provided *In Pursuit of . . . Continuing Research in the Field of the Lincoln Assassination* (New York 1990).

132. For Davis' imprisonment and release, see Davis, *opus cit.*, 640ff.

133. For Lee as college president, see Thomas, *opus cit.*, 376ff.

134. For Lincoln's racial views, see Don E. Fehrenbacher: "Only His Stepchildren," in *Lincoln in Text and Context*, 95–112; G. M. Fredrickson: "A Man Not a Brother: Abraham Lincoln and Racial Equality," *Journal of Southern History*, 41 (February 1975), 39ff.

135. For all the circumstances surrounding the address, and the various texts of it, see Garry Wills: *Lincoln at Gettysburg: the Words that Remade America* (New York 1992).

136. Text in *Speeches and Writings*, ii 555ff.

137. J. B. James: *Framing of the Fourteenth Amendment* (New York 1956); see also H. J. Graham: "Antislavery Background of the Fourteenth Amendment," *Wisconsin Law Review*, 30 (1950), 479ff. See also W. B. Heseltine: *Lincoln's Plan of Reconstruction* (New York 1960).

138. Donald, *opus cit.*, 582–3.

139. G. F. Milton: *The Age of Hate: Andrew Johnson and the Radicals* (New York 1930), 145ff.

140. For the varying positions, see A. O. Craven: *Reconstruction: Ending of the Civil War* (New York 1969) and R. W. Patrick: *Reconstruction of the Nation* (New York 1967).

141. For Johnson's executive action, see J. E. Sefton: *Andrew Johnson and the Uses of Constitutional Power* (New York 1980).

142. For conditions in the South immediately after the end of the Civil War, see J. R. Dennett: *The South As It Is, 1865–66*, ed. H. M. Christman (New York 1965) and J. T. Trowbridge: *Desolate South, 1865–66*, ed. G. Carroll (New York 1966). For the working of the Freedman's Bureau, see Eric Foner: *A Short History of Reconstruction 1863–1877* (New York 1990), esp. 31–2, 64–5, 66, and 111–13.

143. For a recent view of Johnson's handling of Congress, see H. L. Trefousse: *Andrew Johnson: a Biography* (New York 1989).

144. H. M. Hyman: *Radical Republicans and Reconstruction, 1861–70* (New York 1967).

145. For the context of impeachment, see J. E. Sefton: "Impeachment of Andrew Johnson: a Century of Writing," *Civil War History*, 14 (1968), 120ff.

146. David Donald: "Why They Impeached Andrew Johnson," *American Heritage*, 6 (1956), 20ff; Michael Benedict: *The Impeachment and Trial of Andrew Johnson* (New York 1972).

147. J. H. Franklin: "Election of 1868," in Schlesinger and Israel, *opus cit.*, ii; C. H. Coleman: *Election of 1868* (New York 1933).

148. J. Daniels: *Prince of Carpetbaggers* (New York 1958); R. N. Current: *Three Carpetbag Governors* (New York 1967); O. H. Olsen: "Scalawags," *Civil War History*, 12 (1966), 304f. Foner gives a more favorable account of carpetbaggers and Second Reconstruction, *opus cit.*, 129–30, 158, 213, 256.

BOOKS BY PAUL JOHNSON

CREATORS
From Chaucer and Durer to Picasso and Disney
ISBN 978-0-06-093046-2 (paperback)

GEORGE WASHINGTON
The Founding Father
ISBN 978-0-06-075367-2 (paperback)

HEROES
**From Alexander the Great and Julius Caesar
to Churchill and de Gaulle**
ISBN 978-0-06-114317-5 (paperback)

A HISTORY OF THE AMERICAN PEOPLE
ISBN 978-0-06-093034-9 (paperback)

A HISTORY OF THE JEWS
ISBN 978-0-06-091533-9 (paperback)

HUMORISTS
From Hogarth to Noel Coward
ISBN 978-0-06-182591-0 (hardcover)

INTELLECTUALS
From Marx and Tolstoy to Sartre and Chomsky
ISBN 978-0-06-125317-1 (paperback)

MODERN TIMES
The World from the Twenties to the Nineties
ISBN 978-0-06-093550-4 (paperback)

THE QUEST FOR GOD
A Personal Pilgrimage
ISBN 978-0-06-092823-0 (paperback)